Author:
Ann Flagg, M.S. Ed.

Contributing Author:
Debra J. Weinheimer

Illustrator:
Wendy Chang

Editor:
Evan D. Forbes, M.S. Ed.

Editor-in-Chief:
Sharon Coan, M.S. Ed.

Art Director:
Elayne Roberts

Product Manager:
Phil Garcia

Imaging:
Alfred Lau
David Bennett

AF697158

Hands-On Minds-On Science

Our Bodies

Early Childhood

Publishers:
Rachelle Cracchiolo, M.S. Ed.
Mary Dupuy Smith, M.S. Ed.

Teacher Created Materials

Teacher Created Materials, Inc.
P.O. Box 1040
Huntington Beach, CA 92647
ISBN-1-55734-618-6

Table of Contents

Table of Contents *(cont.)*

Introduction

What Is Science?

What is science to children? Is it something that they know is part of their world? Is it a textbook in the classroom? Is it a tadpole changing into a frog, a sprouting seed, a rainy day, a boiling pot, a turning wheel, a pretty rock, or a moonlit sky? Is science fun and filled with wonder and meaning?

Science offers you and your eager students opportunities to explore the world around you and to make connections among the things you experience. The world becomes your classroom, and you, the teacher, a guide.

Science can and should fill children with wonder. It should cause them to be filled with questions and the desire to discover the answers to their questions. And once they have discovered answers, they should be actively seeking new questions to answer.

The books in this series give you and the students in your classroom the opportunity to learn from the whole of your experience—the sights, sounds, smells, tastes, and touches, as well as what you read, write about, and do. This whole-science approach allows you to experience and understand your world as you explore science concepts and skills together.

Why Study the Human Body?

"Yesterday I got a cut on my knee and blood came out. Blood got all over my pants. It was gross!"

"That wasn't blood. It was your guts coming out!"

"Guts can't come out. Guts stay inside of your skin and never come out."

"Well, my guts have come out before!"

Conversations like this one are common in early childhood classrooms all around the world. Children are naturally curious about their bodies and how they work. The book *Our Bodies* utilizes this natural interest to usher children into an age-appropriate study of the human body. Like the children involved in the conversation, you will find the children in your classroom have many misconceptions about their bodies.

After simulating an activity of what "guts" really do, the child changed his idea about "guts." It was discovered that "guts" are a body system designed to break down food. As a result of this activity, a misconception was corrected and the child's view of the human body was greatly expanded.

Scientific Method for Young Learners

What Is Discovery Learning?

Discovery learning is hands-on time for children. The teacher's role is that of facilitator. Instead of using the traditional teacher-has-the-answer format, the teacher becomes the coach, encouraging children to explore and ask "why?" During discovery time, the teacher does not explain the science behind their whys; that comes later in the science lesson (What Scientists Know). It is sometimes difficult for teachers to assume this new role. Teachers have a tendency to want to be explainers or answer-givers to the phenomena that are happening during the discovery experience. During the discovery learning process, children may choose incorrectly or be "way off" on their reasoning about causes or results of a given experience. Realize that this is all right and expected. Do you remember when you were learning to drive a car, start a new sport, or hobby? Many errors were made in the beginning! Only through practice and continual application did you become an achiever in the driving, sport, or hobby. This same process holds true for young learners, too.

There are two types of discovery learning—guided discovery and self-directed discovery. Guided discovery is the actual teacher-guided lesson plan where the teacher sets up experiences with "parameters" (i.e., time, tasks, tools). This type of learning is explained in detail on the following pages.

Self-directed discovery occurs when the teacher removes most boundaries and provides few details or expectations. Children create their own. The most common example of a self-directed discovery approach is a PLEASE TOUCH center where children explore given science topics after initial teacher-directed discovery experiences. In a self-directed discovery setting, children question, explore, and discover on their level and in the direction they wish to go. Language arts is essential to a self-directed discovery experience via reading, writing, listening, and speaking opportunities incorporated into the center area. For example, a child discovers that, ". . . sometimes animals have furry hair and sometimes no hair." The child then draws a picture (including sentences with inventive spelling) to share his/her discovery during a group-sharing "Speak Up For Science" time.

Scientific Method for Young Learners *(cont.)*

What Is the Guided Discovery Lesson Plan?

The guided discovery lesson plan has four parts: the question, the discovery process, the summary of what scientists know, and the real life application. When developing your own science lessons, keep these four parts in mind. Each section builds on the previous section. (This book's activity lessons have been written using this approach.) The question, rather than an educational set or objective, causes children to enter the discovery experience using an open-end (divergent) thinking process rather than a one-answer (convergent) thinking process. What scientists know is only presented after exploration during the discovery experience. Children need to construct their own thoughts about why phenomena happen before the "why" is explained. In the past, children were told the "why" first (usually in form of a factual statement or definition). They were then asked to complete an experiment based on that definition. That experiment, in essence, contained no true discovery element because it was simply a hands-on definition of what the teacher had initially stated to children during the onset of the science lesson. The science simulation is of extreme importance. If teachers do not allow young learners to kinesthetically assimilate what has been taught, the child will lose a majority of information given. The instant application of newly gained science knowledge will help children remember better and be able to apply this new knowledge more effectively.

Question

The easiest way to come up with the question is to make a statement or look at the overall objective and switch wording around. For example, "Predict how many colors a dog can see" is far less open-ended than, "How many colors do you think a dog can see?"

Discovery Experience

This is the time when children hypothesize through predictions (P), investigations (I), and draw conclusions (DC). Whenever possible, create discovery experiences that cause children to move through the P-I-DC process more than once. This repetitive process helps children understand and appreciate the science motto, "Scientists are always changing their minds." For example, if children are asked to discover which animals live under, on, and above ground, they will eventually discover that some kinds of like animals (spiders) live in all three areas instead of one area as some children may have previously thought.

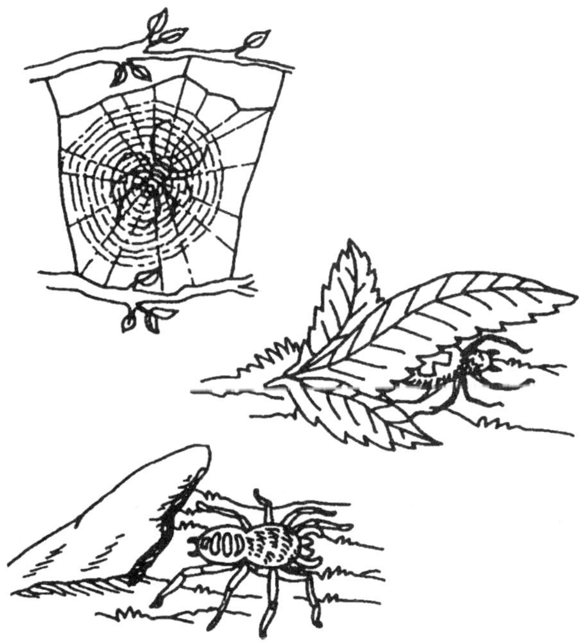

Scientific Method for Young Learners *(cont.)*

What Is the Guided Discovery Lesson Plan?

What Scientists Know

The most crucial component to this part is teacher attitude. Instead of saying, "This is what I know about. . ." or "Here are the facts about. . .," try saying, "This is what the scientists know about. . ." The reason for this is two-fold. One, the teacher conveys to children that he/she is not the science answer-man/woman. The teacher is simply sharing what scientists know and what the teacher has learned, too. Two, if a student asks a question that cannot be answered by the teacher, no panic buttons need to be pressed! The teacher simply says something like this: "That is an excellent question. I don't know what the scientists know about that, but I'll be glad to try and find the answer for you and tell you what I've learned."

Note: Explanations within "What Scientists Know" in *Hands-On Minds-On Science: Our Bodies* have been written using "simple text" so young children can understand otherwise complex concepts.

Real Life Application

Real life application gives children a chance to internalize their new knowledge. For example, suppose a teacher has taught what the brain can do for each one of us. A demonstration in the classroom will help the students understand this, of course. A real life application, however, in which each child plays a game of Brain Concentration will verify the understanding and solidify the retention of the new knowledge. (See *Brain in Charge*, page 42.)

Science-Process Skills for Young Learners

Observing

All information is acquired through the process of observing. That makes this skill the most fundamental of all the process skills. Children have been making observations all their lives, but they need to be made aware of how they can use their senses (seeing, hearing, smelling, tasting, touching) and their prior knowledge (life experiences) to gain as much information as possible from each experience. Teachers can develop this skill in children by asking them questions such as these:

Can you tell me what you see? What did you hear?

Can you point to what you are talking about?

Communicating

Humans have developed the ability to use language and symbols which allow them to communicate not only in the "here and now" but over time and space as well. The accumulation of knowledge in science, as in other fields, is due to this process skill. Of the two main forms of communication, written and oral, young children have few problems with the oral. The written, however, often poses problems for teachers. Please keep in mind that "a picture speaks a thousand words." In an active-learning science environment, this process skill will include many visual displays (drawings and models), inventive spelling, and much talking, sharing, and questioning. The style of questioning for the teacher should encourage children toward expressing their thoughts:

Can you draw a picture of what you just saw?

May I join your team and talk about what we just saw happen?

Can you write a sentence to go with the picture you just drew?

Classifying *(Comparing, Ordering, Categorizing)*

Once children are actively engaged in observing and communicating, they naturally begin to make comparisons, try to find order, and categorize what they are observing or doing. Comparing means noticing the similarities and the differences. Ordering includes finding patterns (smooth to rough, bright to dim, few to many) and sequencing events (time line or cycle). Categorizing includes trying to find connections among objects based on logical rationale (color, size, weight, etc.). Questions for encouraging this process skill would include the following:

How are these items alike? How are these objects different?

What happened first/second/last? What will happen next?

Can you find a way to show these objects are alike, but also different?

Can you show me three different groups you can make by using the same objects for all of your groups?

Do-Along Books

For each of the three body sections (Why Is My Body Special? Who Is the Boss of This Body? and "Every Body" Has Five Senses) there is a supplementary do-along book. Each book is designed so that children, in a hands-on/kinesthetic way, can create concrete images that illustrate concepts taught. Each book consists of eight pages (four large pages, cut out and folded in half) containing simple sentences that correspond to the do-along book's activity page. The reason art work may not necessarily be "complete" on some pages is so children become more involved in the book's creation. On each page immediately following "Just the Facts," you will find specific directions for what is needed, as well as directions for completing that section's do-along book. For the general assembly of a do-along book, follow the directions below:

1. Copy all large do-along pages; cut around outer rectangle's black line.

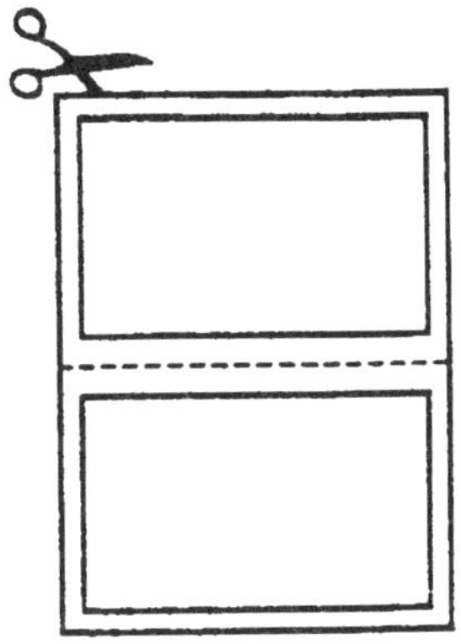

2. Fold pages in half on dashed line. Glue undersides of "pages" together; allow to dry.

3. Layer book in order: front cover and page 1, pages 2 and 3, pages 4 and 5, etc.

4. Make certain all left-side edges are flush. Staple book together near top, middle, and bottom of left-side edge. (Optional: Use hole puncher/string method.)

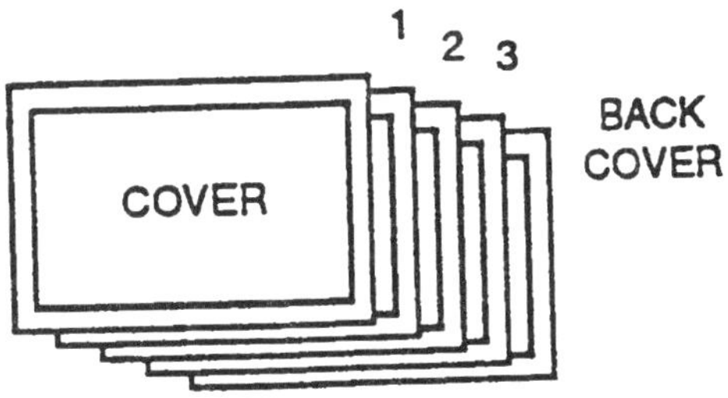

5. Write "title" of book on cover, along with child's name (or have the child write own name). Have each child decorate the front cover after overall concepts (pages) have been introduced.

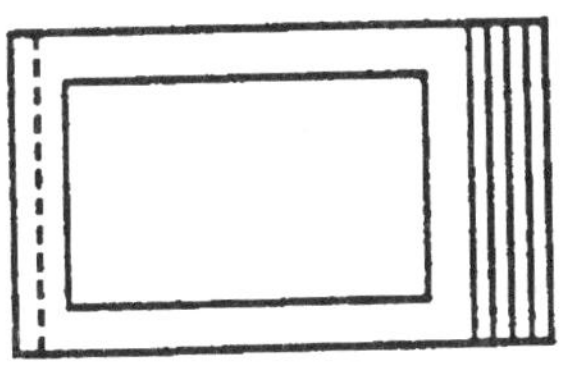

Just the Facts

This section of *Our Bodies* is designed to familiarize children with one of their favorite topics—themselves. Self-esteem will increase as each child creates a life-sized shape book which represents his/her own body in form and size. As they work through the activities, your children will learn terminology they can use to communicate about the human body. They will discover that each human body has the same basic design, but that there are many variations in facial features as well as body size and shape. In this section your children will learn the following:

My body is just like yours in many ways.

Despite apparent similarities, there is nobody with a body exactly like mine.

My body can do many things.

Each outside body part is designed to perform specific tasks.

Words like back and front and inside and outside can be used to describe my body.

Body vocabulary is necessary if I want to communicate about my body.

Every person in my class is gifted in some area. Every person is special.

We can use our bodies to help others and to do kind things. This can make our classroom a nice place to be.

My Body Is Amazing

Supplies needed for this do-along book:

- stamp pads
- pencils
- crayons
- scissors
- glue
- tagboard

Cover: Your hand print is unique.
Press each child's left hand on a well-inked stamp pad. Then, place each child's hand within the outline of the hand on the cover.

Say, "Look! Your hand is just like this one. Look at all the lines on your hand. They are different from everyone else's."

Page 1: Collect thumb print autographs.

Provide children with well-inked stamp pads. Have each child rub his/her thumb on the stamp pad and then press firmly to make his/her thumb-print in other students' do-along books. Write the child's name by his/her thumb print.

Page 2: Draw a person.

On a large sheet of paper, have volunteers help you draw a person. Encourage your children to add more details and draw a more complete body than they could have when they were younger.

Provide page 2 of the do-along book. Ask children to work independently to draw their bodies. Encourage them to look at their body outlines and think about the class drawing in order to add as much detail as they can.

Page 3: A Silly Sammy puzzle.

Copy page 15 for each child. Detach the top portion. Instruct children to color the pieces, cut them out, and glue Silly Sammy back together.

Page 4: Bragging about my body.

On this page, have children draw or write some things they can do.

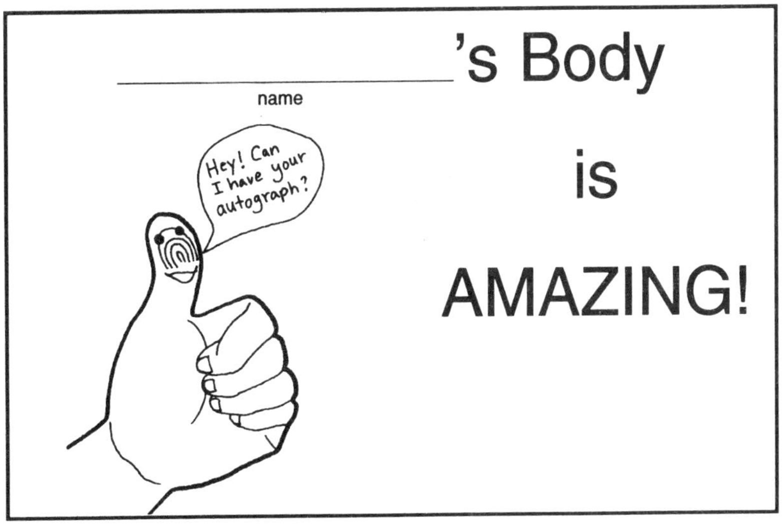

Page 5: Add a body picture that is taller than the book.

For this do-along page you will need the bottom part of page 15 for each child. Glue this paper to tagboard and cut the body out. Have each child decorate the front and back of the body with crayons. Glue the body over the X on page 5 of the do-along book so that the hands, arms, and head stick out above the book pages and the feet rest on the grass.

My Body Is Amazing (cont.)

1

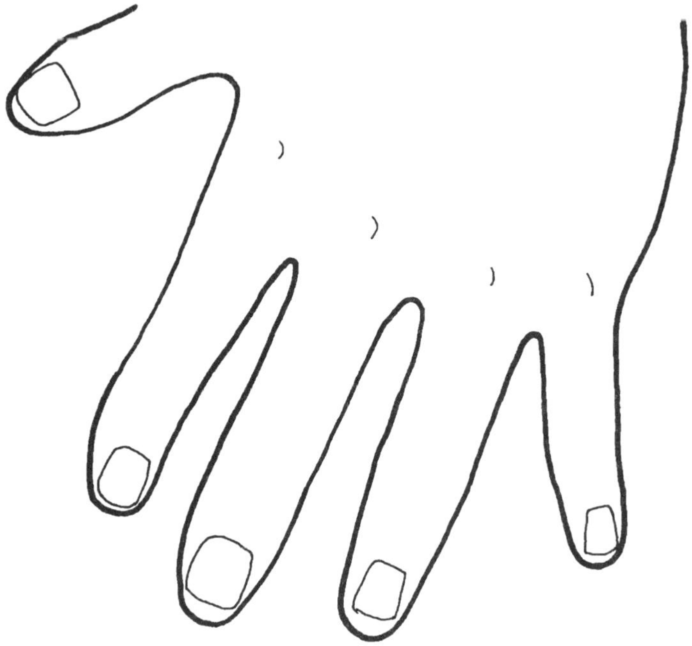

We are each special!

____________________'s Body
name

is

AMAZING!

My Body Is Amazing *(cont.)*

3

Help Silly Sammy "Get It Together"!

Draw your amazing body.

2

My Body Is Amazing *(cont.)*

5

I have a FRONT and a BACK.

This is What I Can Do

4

My Body Is Amazing (cont.)

Use with page 3 of the *My Body Is Amazing* Do-Along Book.

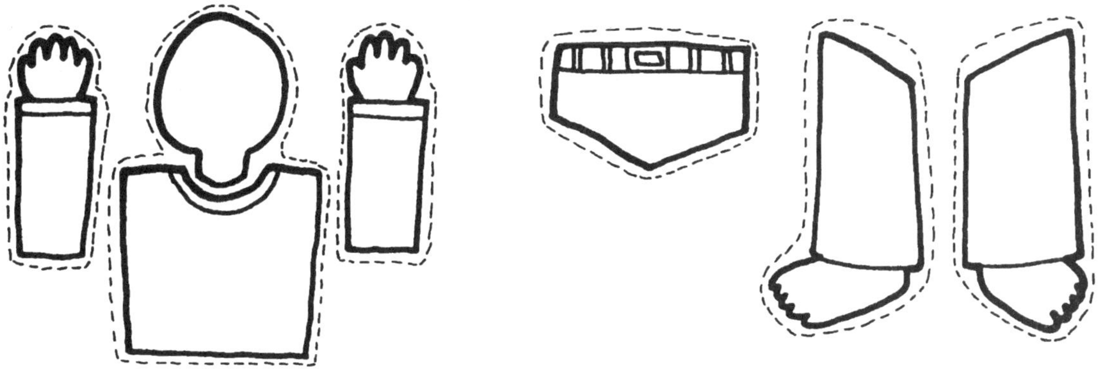

Use with page 5 of the *My Body Is Amazing* Do-Along Book.

You Look Like Me

Science Concept: My body is like your body.

Concept Literature: *Bein' with You This Way* by W. Nikola-Lisa

Question

How is my body the same as other children's bodies?

Materials Needed for Each Child

- one sheet of butcher paper folded in half (This should be large enough to fit a child's entire body on each side of the folded sheet.) A standard roll of 4'(120 cm) paper works well.
- wide-tipped dark marker

Discovery Experience

Over the next three lessons, each child will construct a life-sized model of the outside of his/her body. In later activities, your children will open the models to add their internal organs and systems. For further instructions about this model, see Body Builder Card 1, page 88.

1. Use a pencil to trace the outline of each child's body onto the folded butcher paper. Trace around the child's hair to help individualize each tracing. The fold will be a part of the spine of a life-sized body book, so be sure that each child's right arm and shoulder line up against the fold of the paper as shown. Remember to write the child's name on his/her paper. Note: Most 5-to-8-year-old children will fit within the parameters of a large piece of butcher paper. To accommodate a tall child, allow his/her feet and ankles to hang off the butcher paper and add feet extensions. To accentuate the pencil lines, trace over the outline with a wide-tip marker.
2. Display all the body outlines. Keep the body outlines folded.
3. Give your children a few moments to be detectives and discover how all their body outlines are alike. Please note: If you have a physically challenged child in your classroom, use this as an opportunity to affirm his/her uniqueness.
4. As your children share their findings, make a class chart like the one shown.

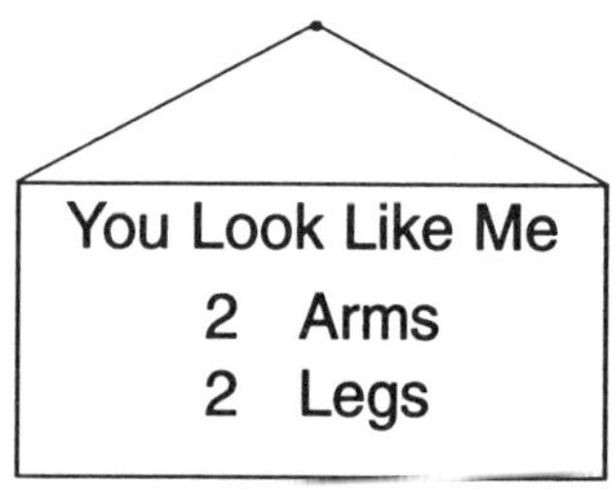

What Scientists Know

Although human bodies have similarities, they are not all the same. Most of us have two arms, two legs, two hands, two feet, ten fingers, and ten toes. However, there are many characteristics that make us different.

Real Life Application

In many ways, our bodies reflect each other. Use the mirroring activity below to extend today's experience. Group your children in pairs and have them sit facing one another. Lead them through this dialogue of discovery:

"Pretend your partner is a mirror. Look into the mirror and see how many eyes you have."

"Now feel your own face. How many eyes do you have?"

"You and your partner have the same number of eyes."

Use the same type of dialogue to ask about the mouth, ears, nose, and fingers. Add children's new ideas about similarities to your class chart.

Complete the cover of the *My Body Is Amazing* Do-Along Book.

No Body Is Like Me

Science Concept: Although we may be similar, every body is special and unique.

Concept Literature: *Why Am I Different?* by Norma Simon

Question

What makes me special and unique?

Materials Needed for Each Child

- traced body outline from the previous activity
- colored markers or crayons
- a mirror for every two children

Discovery Experience

1. Group your children into pairs. Invite each pair to sit facing each other.
2. Guide them through this dialogue and allow time so they may share the answers with their partners:

   -What color are your partner's eyes?

   -What color is your partner's hair? Is it long or short? Curly or straight?

   -How would you describe your partner's skin color? Does he/she have freckles?

   -What color is your partner's shirt (dress, pants, etc.)?

   -Does your partner wear glasses?

   -How are you and your partner different?

3. Ask your children to keep all of these special characteristics in mind so they can paint realistic features on their body outlines.
4. Place the outlines on the floor or any large area where your children can color on them. Supervise this closely so your children color only on the front of their body outlines.
5. Have your children use their mirrors to look at themselves while they paint their bodies. Their pictures should look as much like them as possible.
6. Cut out each child's body outline. Remember that you want the body outline to open like a book, just as in the previous activity. Allow the fold to remain intact at the right arm and shoulder and cut through both pieces of paper. If you are working with older children, have them cut out the outlines themselves.

No Body Is Like Me *(cont.)*

What Scientists Know

Most human bodies have the same parts, but those parts are amazingly different. No one is quite like you. People in the same family, even twins, have unique bodies. Many attributes make you different from anyone else. What colors are your skin, your hair, and your eyes? Are you tall or short, fat or thin, or are you somewhere in the middle? Even the shape of your nose belongs just to you. No body is exactly like you.

Real Life Application

May I have your autograph, please? Children can give their autographs in more than one way. Continue to celebrate each child's uniqueness by completing page 1 of the *My Body Is Amazing* Do-Along Book. Provide children with well-inked stamp pads and invite them to collect classmates' thumb prints. Each child may then write his/her name next to his/her thumb print.

As an extension, have your children take home and complete Different and the Same found on page 19. This take-home assignment will help children review the concepts of same and different as they compare themselves with people in their families.

This lesson and the next are repeated on Body Builder Card 2.

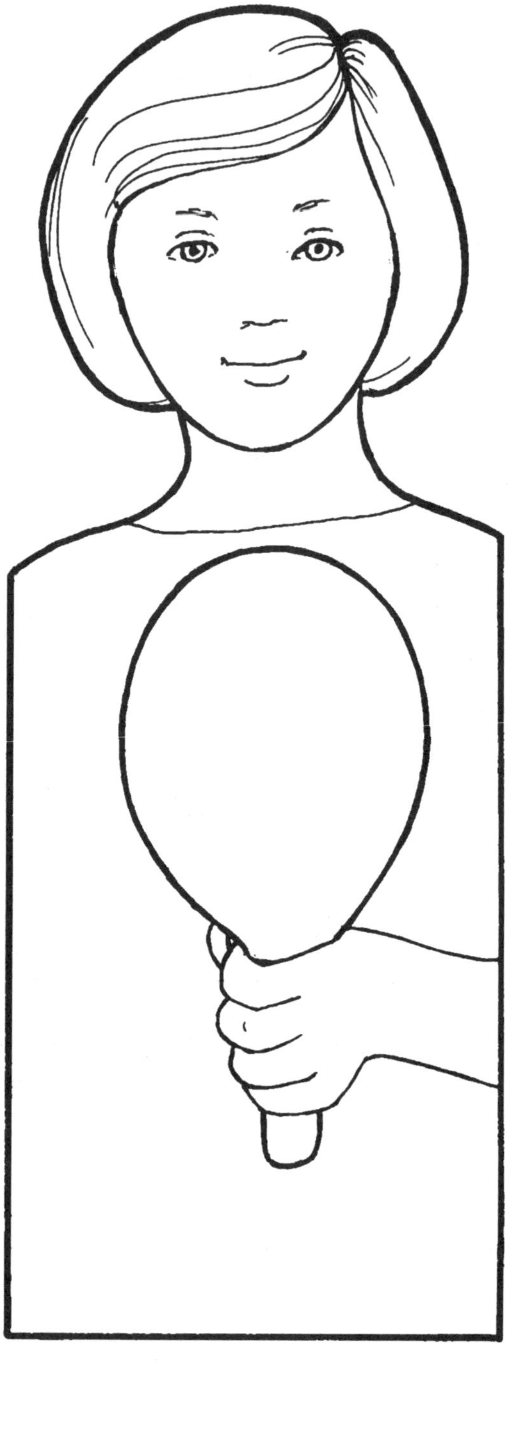

Different and the Same

Directions:

1. Ask one of your parents to help you.
2. Draw a picture of your face and your parent's face.
3. List several ways you are the same and different from your parent.

This is Me	**This Is ______________!**

Ways We Are The Same:	Ways We Are Different:
______________________	______________________
______________________	______________________
______________________	______________________
______________________	______________________
______________________	______________________

Back to Back

Science Concept: I have a front and a back.

Concept Literature: *Here a Chick, There a Chick* by Bruce McMillan and *Exactly the Opposite* by Tana Hoban

Question

How does my back look and feel?

Materials Needed for Each Child

- traced body outline with front colored
- crayons

Discovery Experience

1. Place body outlines face down on the floor or on any large area where children can paint. The uncolored back of the body outline should be face up.
2. Show the front and back of several items, like a book or a box of cereal, and then begin this dialogue: We each have a front and a back. You have already colored the front of your body models. On the front of your body, you can find your face. Point to your front side. Now point to your back side. How is your front side different from your back side?
3. Have your students stand in a circle in your classroom. Arrange them so they all stand facing the same direction and each child can view another child's back.
4. Lead your children through the following dialogue:
   "Do you see a face when you look at your friend's back?"
   "Do you see his/her hair?"
   "Can you find a nose (eyes, mouth) buried underneath the hair?"
   "Do you see his/her arms?"
   "Do you see his/her back?"
   "Do you see his/her legs?"
5. Ask each child to look over his/her shoulder to view his/her own back.
6. Direct your children to their body outlines and supervise them as they color the backs of their bodies.

What Scientists Know

Your back is an important part of your body. It is the posterior part, which means it faces backward. Without this side of your body you would not be able to lie on your back, lean back in a chair, or sit down. Besides, you would look pretty silly walking around without a back for your front.

Real Life Application

What does your back feel like? Have your children sit cross-legged on the floor in a large circle so that they are facing another child's back. Everyone should be facing the same direction. Allow your children to gently feel the shoulders, backbones, and waists of their partners. Next, demonstrate a group back rub. Invite the "back" child to rub the back in front of him/her. After a while, switch directions and repeat the whole process. Every child will get a chance to be in "front" and in "back."

Complete page 5 in the *My Body Is Amazing* Do-Along Book.

Cut It Out

Science Concept: My body has many parts with special names.

Concept Literature: *Here Are My Hands* by Bill Martin, Jr., and John Archambault

Question

What are the names of my outside body parts?

Materials Needed to Make a Class Big Book

- *Here Are My Hands* by Bill Martin, Jr., and John Archambault
- labels from page 23
- 5 large, white pieces of construction paper suitable for making a class big book

Materials Needed for Each Child

- magazine that can be cut up
- scissors
- glue

Discovery Experience

1. To prepare your class big book, cut out and glue each label from page 23 to a separate piece of construction paper. Bind the pages with staples, yarn, or loose-leaf rings.
2. Read *Here Are My Hands* aloud with your children.
3. Read the story a second time. This time, ask children to point to their body parts as you read about them in the book.
4. Show the blank big book to your class and ask them what is missing. Tell the class that they will be adding the pictures to the book.
5. Pass out a magazine and a pair of scissors to each child. Ask your children to find pictures in the magazines that show each body part identified on the labels. Have each child cut and glue his/her pictures to the appropriate page in the big book. With younger children, look for each body part as a class. For example, "Everyone find a face in his/her magazine. Cut it out and paste it on the page that says, Here is my face. Look at my eyes . . ."
6. Share the completed book with your class and keep it available for your children to read again and again.

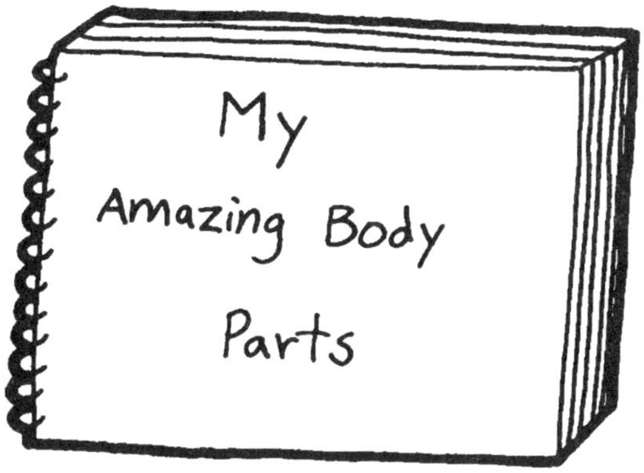

Cut It Out *(cont.)*

What Scientists Know

Your body has a head, a torso (the mid-section of your body), two arms, and two legs. We can be more specific about describing or explaining our bodies by using words that name the specific parts; words like neck, knees, hands, and toes are body words too.

Real Life Application

Read *Here Are My Hands* aloud and record it on a cassette tape. Encourage acting out of the story during play time or as part of a center. Children's tape players with big buttons work best for children to operate. Children may wish to expand the story, as well. What else can my hands do besides catching and throwing? In addition, have children find the parts of the body on their body outlines to enable an easier connection between their real bodies and their abstract outline bodies.

Complete page 2 in the *My Body Is Amazing* Do-Along Book.

Cut It Out *(cont.)*

My Amazing Body Parts
Here are my hands and arms. I can carry and lift.
Here are my feet and legs. I can run and kick.
Here is my face. Look at my eyes, ears, nose, and mouth.
Here is my body. It is amazing.

Label It

Science Concept: My body has many parts that are connected.

Concept Literature: *Goodness Gracious!* by Phil Cummings

Questions

What are the names of my outside body parts? Where does each part belong?

Materials Needed for Each Child

- head, face parts, and body labels, page 26
- body and legs, page 27
- magnetic strips or Velcro
- glue
- masking tape

Discovery Experience

Before you begin, cut out the pieces of Silly Sammy found on pages 26–27. Color in his clothes and face parts. Laminate Sammy's parts to make them durable. Cut off the arms and legs at the dash marks. Use tape to reassemble the body and face as shown in the example. Attach small pieces of magnetic or Velcro strips to all the body labels and same size magnetic or Velcro pieces to the corresponding locations on the body outline parts.

1. Display the unlabeled, mixed-up Silly Sammy. Introduce him as Silly Sammy and tell your children this funny story designed to provide practice with naming body parts.

   "Silly Sammy woke up this morning and did not feel quite right. He hopped out of bed and walked on his hands into the bathroom. "I feel strange," said Silly Sammy. "I wonder what is wrong with me?" Do you know what is wrong with Silly Sammy? Well, Sammy took one look in the mirror and said:

   "I have feet instead of hands."
   "My arms are where my legs should be."
   "My eyes are on my chin."
   "I can't even see."
   "My ears are all mixed up."
   "I am as silly as can be."
   "My mouth is a bit too high."
   "Please help me!"

   Can you help Silly Sammy to get his body straightened out?

2. Unstick each of the body pieces and let your children tape them back into their correct positions.

3. Tell your children you hope this never happens to Silly Sammy again! But in case it does, let's label and name Sammy's body parts so he can put himself back together next time!

4. Help your children to label each body part.

5. Keep Silly Sammy and the labels available for children to manipulate and review.

Label It *(cont.)*

What Scientists Know

Each body part has a job. To do its job well it has to be in its proper place on your body. Your hands would become very tired if you had to walk around on them all day, but your feet do the job just fine. It would be tricky to brush your teeth with your foot holding a toothbrush, but your hand can brush really well. Your body is designed to help you get the job done.

Real Life Application

You may be familiar with the song *Head and Shoulders, Knees and Toes:*

Head and shoulders, knees and toes, knees and toes
Head and shoulders, knees and toes, knees and toes
Eyes and ears and mouth and nose
Head and shoulders, knees and toes, knees and toes!

Invite your students to sing this song and point to their body parts as they sing their names. Then add this verse:

Fingers, hands, and elbows and elbows
Fingers, hands, and elbows and elbows
And legs, knees, feet, don't forget about my toes!
Fingers, hands, and elbows and elbows

Complete page 3 in the *My Body Is Amazing* Do-Along Book.

Label It *(cont.)*

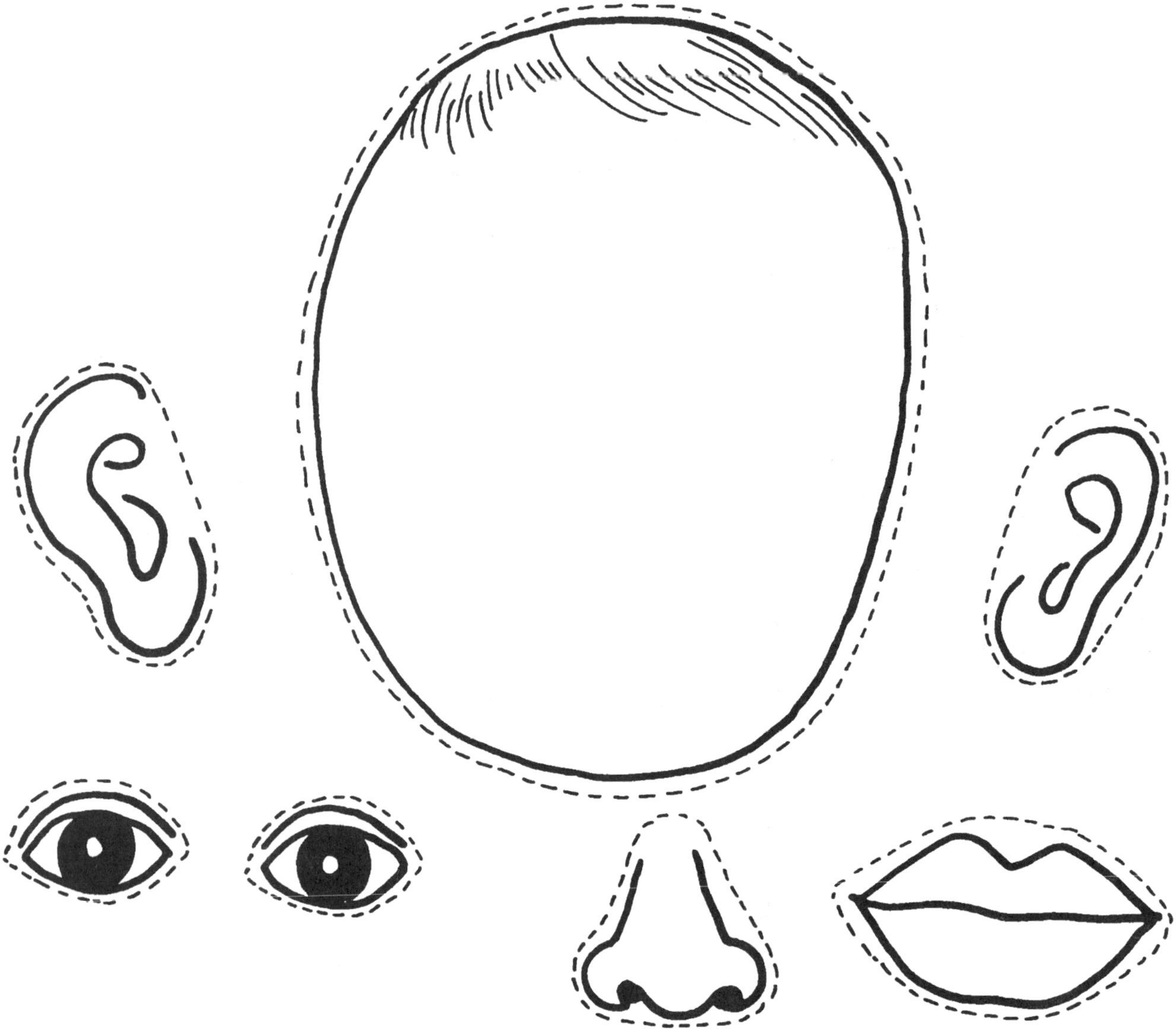

Body Labels (for Silly Sammy)			
Leg		Chest	
Arm		Face	
Stomach			

Label It *(cont.)*

Can You Do That?

Science Concept: My body can do amazing things.

Concept Literature: *Rainbow Fish* by Marcus Pfister

Question

What unique things can I do?

Materials Needed for Each Child

- body outline (entire outside should be completed)
- blank piece of paper
- colored markers or crayons
- glue
- ball

Discovery Experience

1. Have your children sit in a circle on the floor. Begin a ball-rolling game by telling your children something you can do with your body. For example, "I can hug my children." Roll the ball to one child at a time. When a child has the ball, ask him/her to tell the group one thing he/she can do. When the child is finished, instruct him/her to roll the ball back to you. You then roll the ball to another child and continue until every child has had a turn. (Older children may enjoy tossing and catching a ball, instead of rolling it.)
2. Have children think of something amazing they each can do. Then have them write those things on a blank piece of paper, or you can take their dictations. Glue these sheets of paper on the left inside sheets of the body outlines. Children may then draw pictures of things they can do well around the writing paper on the inside covers.
3. Encourage children to share the amazing things that their bodies can do.

What Scientists Know

Your body is more amazing than the fanciest machine or the best computer. You can do lots of things. You can think, laugh, cry, and sneeze. You can jump rope, run, sing, dance, and clap your hands. Your body is always working, even when you are sleeping. You can do amazing things.

Real Life Application

Remind your children that each of them can do some things very well, but no one can do everything perfectly. Share *Rainbow Fish* by Marcus Pfister. Have children brainstorm a list of things they can do to help others in their class, home, or neighborhood.

Complete page 4 in the *My Body Is Amazing* Do-Along Book.

Just the Facts

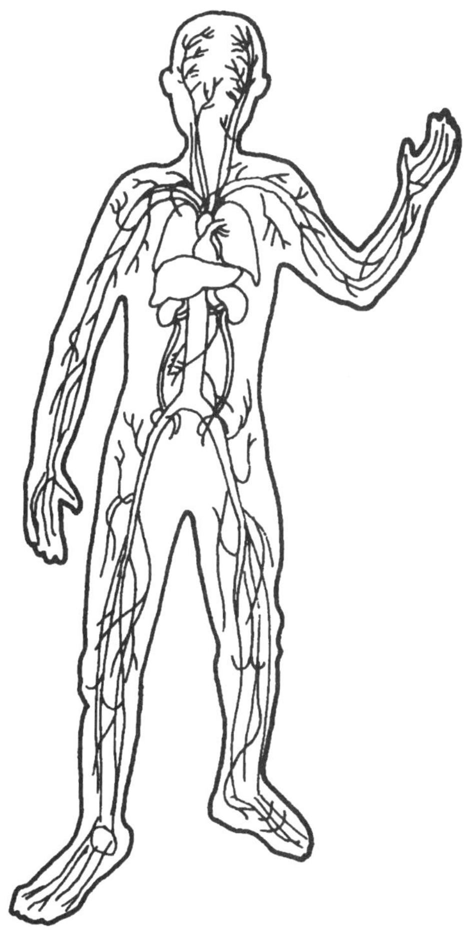

This section of *Our Bodies* will introduce the children in your class to the inside systems of their bodies, with a special emphasis placed on the brain. Transition activities like "The Inside and Outside of Me" will present the concept that there is more to the human body than what is seen on the outside. From there, activities like "Popping with the Brain" and "Brain in Charge" will establish the brain as the "boss of the body," but emphasize that the brain needs the five senses to do its job. In fact, your children will learn that the five senses act like detectives for the brain in that they gather information from the outside to help the brain think and function as the boss. The nervous system is introduced next as a communication network between the brain and the five senses and the brain and body. The brain is introduced before any of the other body functions because many of the other activities will refer to the brain.

Teacher Information:

Note: You will find information about the brain and nervous system written at a child's level after each activity.

The brain is an extremely complex and chemically active organ. Therefore, scientists have only a partial knowledge of the brain and how it functions.

The brain is the center of all mental functions. It has automatic functions like controlling the heart's beat, the digestive process, and conscious functions like recalling information and walking.

The brain has three main structural parts: the *cerebellum*, the *cerebrum,* and the *brain stem.* The cerebrum is the thought center, the cerebellum coordinates muscle movement, and the brain stem controls automatic and life sustaining functions like breathing.

Your five senses are continually gathering information. The brain screens it all, directs your body to attend to stimuli that are important, and tosses out information that is not important.

The nervous system branches out and reaches every part of the body. The system has two main sections: the central nervous system, which consists of the brain and spinal cord, and the peripheral nervous system, which consists of the rest of the nerves which branch out all over the body and communicate with the brain via the spinal cord.

Nerves are really bundles of nerve cells that carry electrical messages. Nerves can be as thick as rope or as thin as a hair.

The ears and eyes are joined directly to the brain by cranial nerves. There are 12 cranial nerves in all that join the brain to important parts of the head.

The Inside and Outside of Me

(For general book-making instructions, see page 9.)

Supplies needed for this do-along book:

- snapshot or copy of snapshot of each child
- glue
- scissors
- crayons or colored markers
- finger paint
- paper
- Jello powder
- letter-size envelope

Cover: Paste on the cover a snapshot or draw a copy of a snapshot, or use a child's self-portrait.

Page 1: Open up a body.

After completing the activity *The Inside and Outside of Me*, provide your children with glue and scissors. Have your children cut out the Outside Body from page 35. Glue the tab on the square marked "glue" on page 1 of the do-along book. Have children color and personalize their pictures.

Page 2: Add wrinkles to my brain.

It is not how big your brain is, it is how many wrinkles you have. Scientists who have studied the brains of very intelligent men and women have found that intelligent people have more wrinkles than people with average intelligence. Show your children how to use a pencil to add wrinkles to the brain found on page 2.

Page 3: Make a scratch and sniff spot.

Use finger-paint paper and cut out a circle to match the circle on page 3. With the shiny side up, sprinkle a bit of clean water on each child's finger-painted paper circle. Say, "I have some mystery powder. Can you figure out what it is?" Give each child a spoonful of Jello powder. Have them sprinkle most of it on the wet circle and then taste a bit of what is left over in their hands. Ask, "Using your five senses, can you figure out what my mystery powder is?" (Jello). Let the circles dry and then glue the bumpy, Jello-powdered circles on page 3 of the do-along book. Your children can continue to sniff and feel the circle each time they read their do-along books.

Page 4: Puzzle the brain.

Cut out the brain puzzle pieces on page 36. Give each child a set of the three pieces. Use the dialogue from *Brain Puzzle*, page 43, as you put the puzzle together.

Page 5: Read a message.

Glue a letter-sized envelope onto page 5 of each child's do-along book in this way: Lick the flap on the envelope, but tuck it inside and seal. Place glue along the three edges, but when you get to the top place glue only on the back V. The back will stick securely, and you can still put your hand in the envelope. Duplicate and cut out the message on page 35 for each child. Your children may trace over the dotted lines, read the note, and tuck it inside the envelope.

Page 6 and 7:

Read the imaginary dialogue between the brain and the nerves in the hand.

The Inside and Outside of Me *(cont.)*

1

There is an **inside**
and an **outside**
of me.

There are maky parts
inside me that
I cannot see

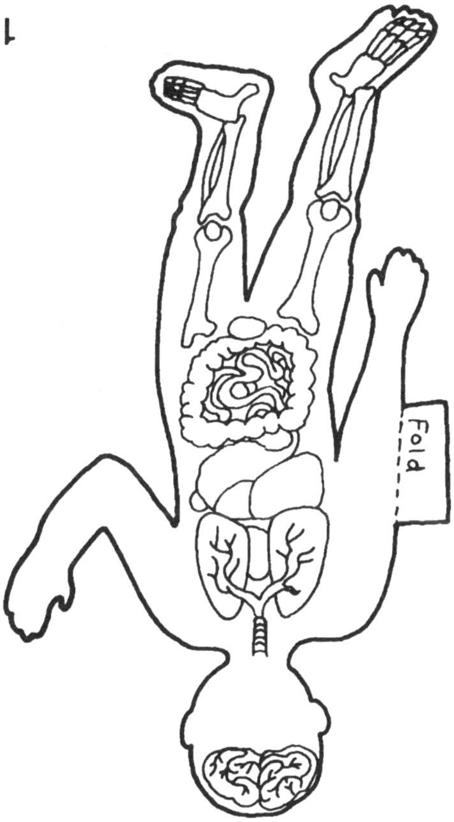

The Inside And Outside Of Me

by ______________________________
child's name

The Inside and Outside of Me *(cont.)*

My five senses help my brain. 3

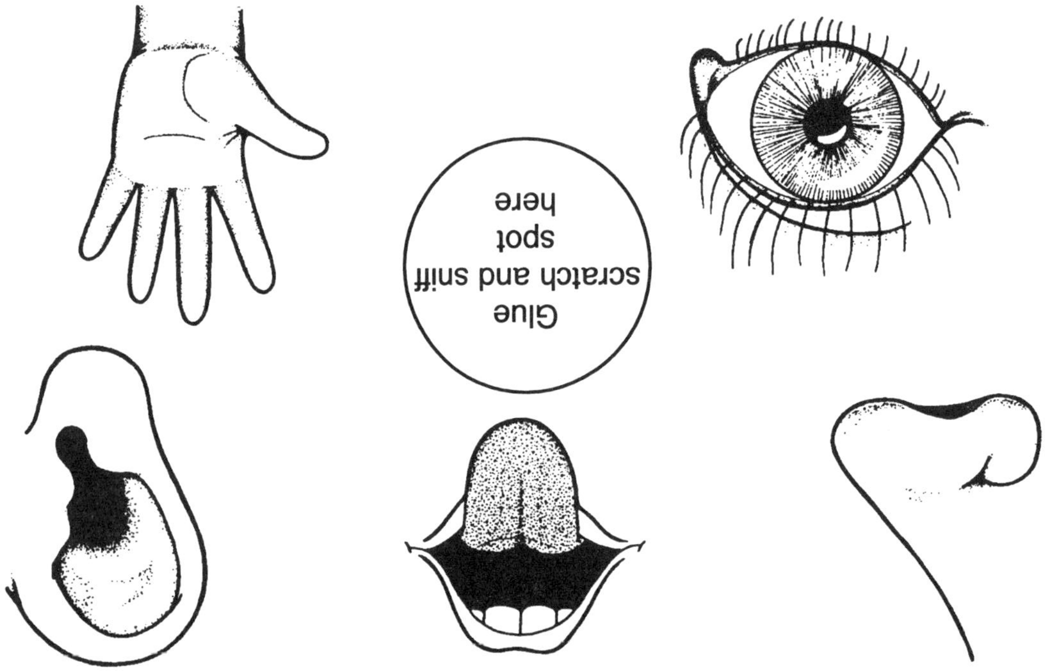

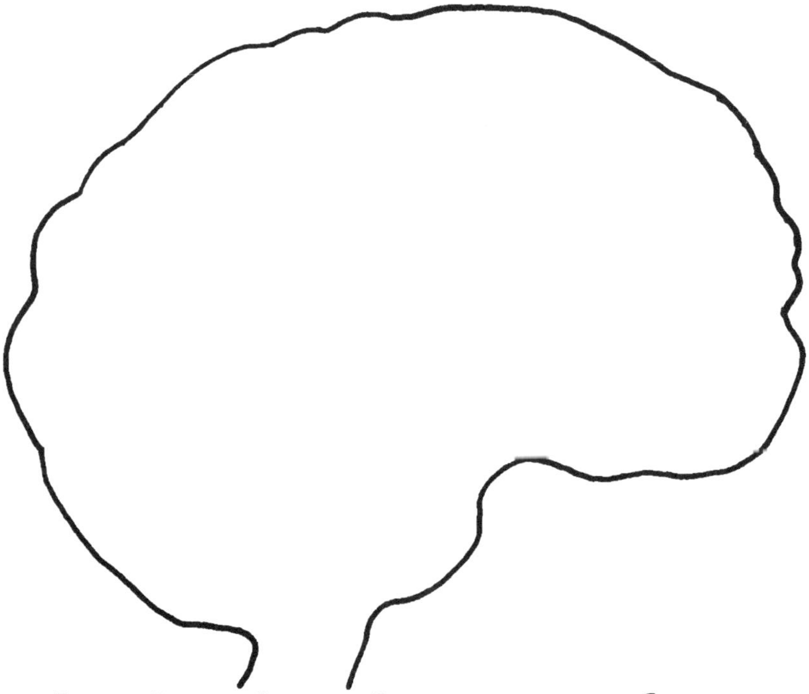

My **brain** is the **boss** of my body. 2

The Inside and Outside of Me *(cont.)*

My nerves help my brain and body work together to get the job done. 5

Glue envelope here

Glue envelope here

Glue envelope here

From:
Brain

To: Hands

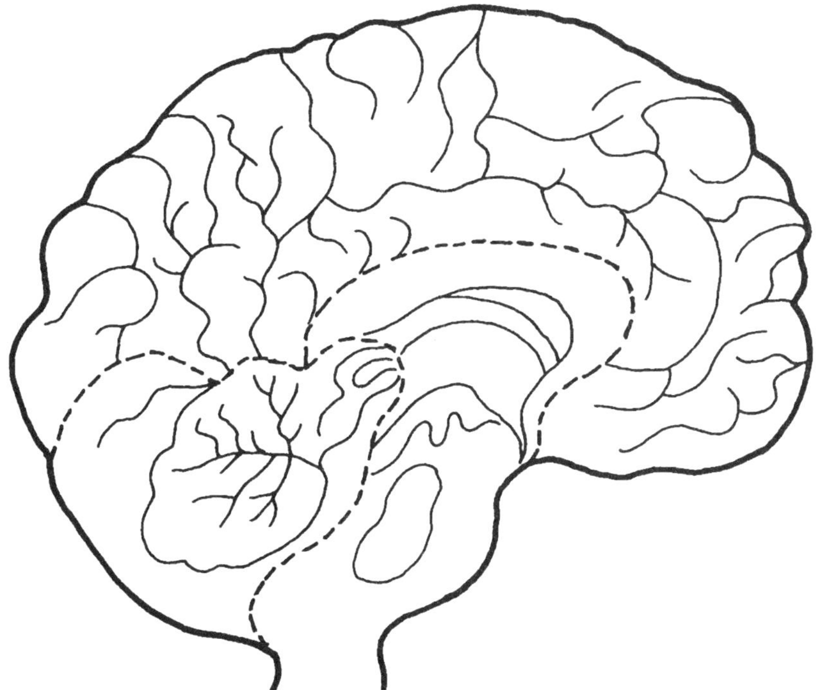

Each part of my brain has a different job to do. 4

The Inside and Outside of Me *(cont.)*

7

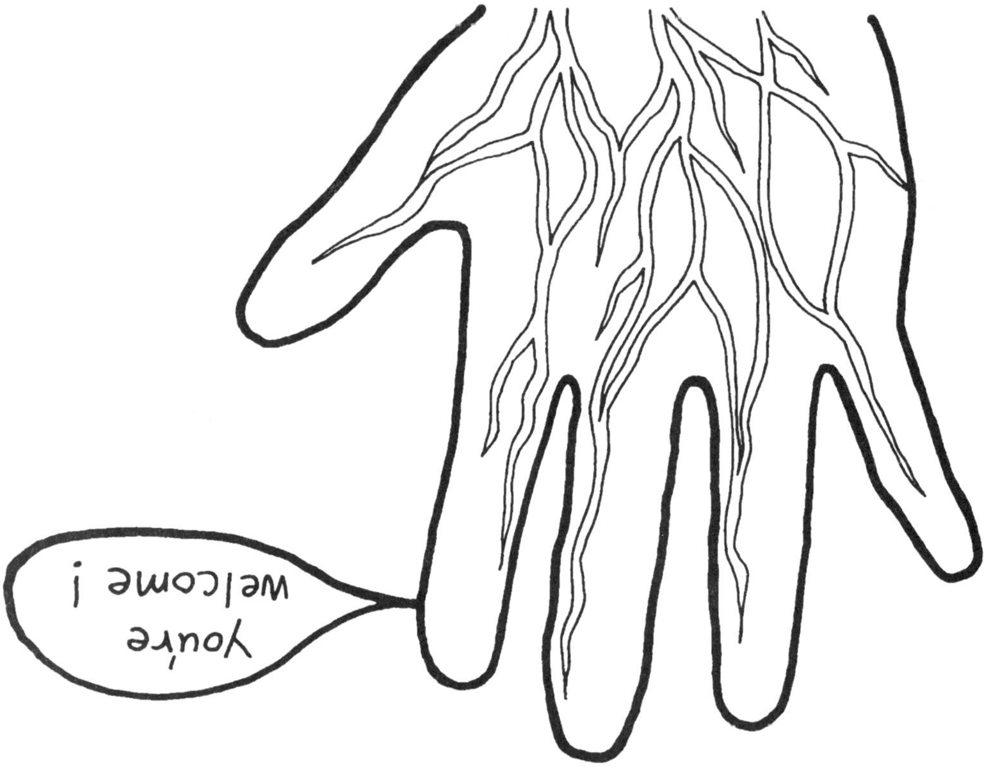

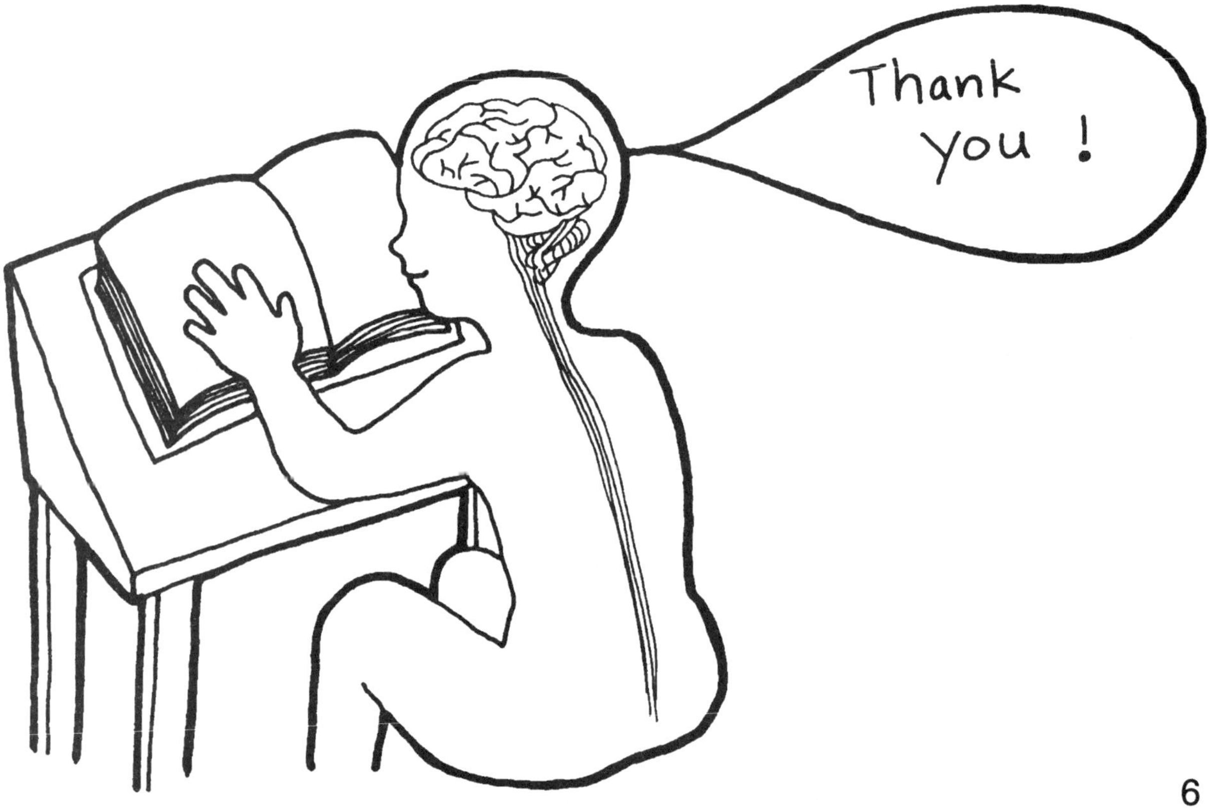

6

The Inside and Outside of Me *(cont.)*

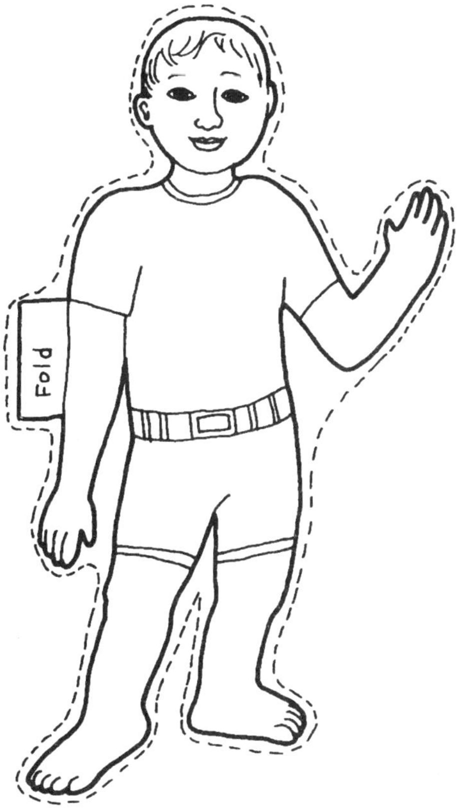

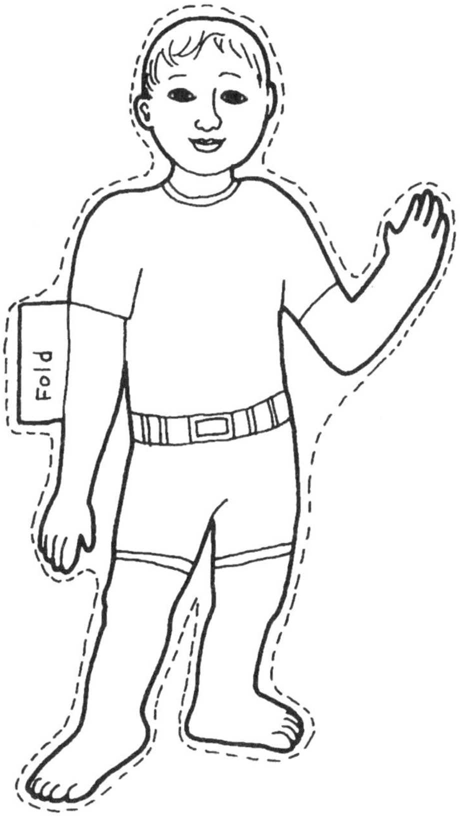

Dear Hand,

Turn the page.

Sincerely,

Brain

The Inside and Outside of Me *(cont.)*

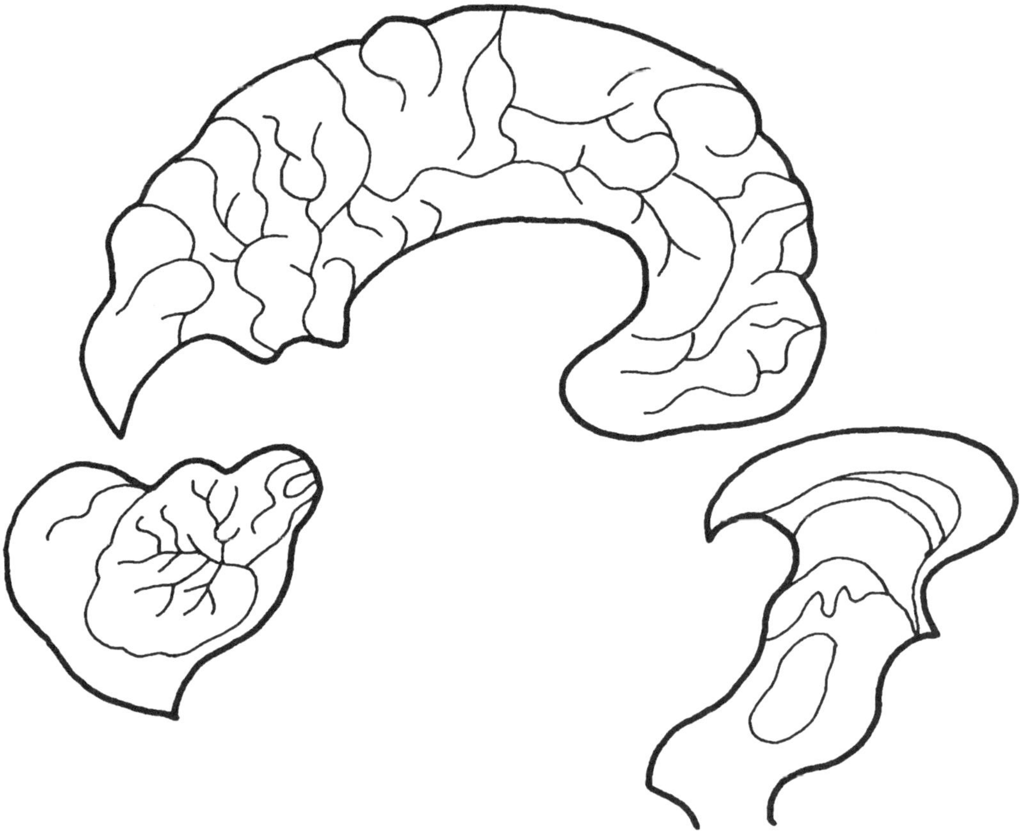

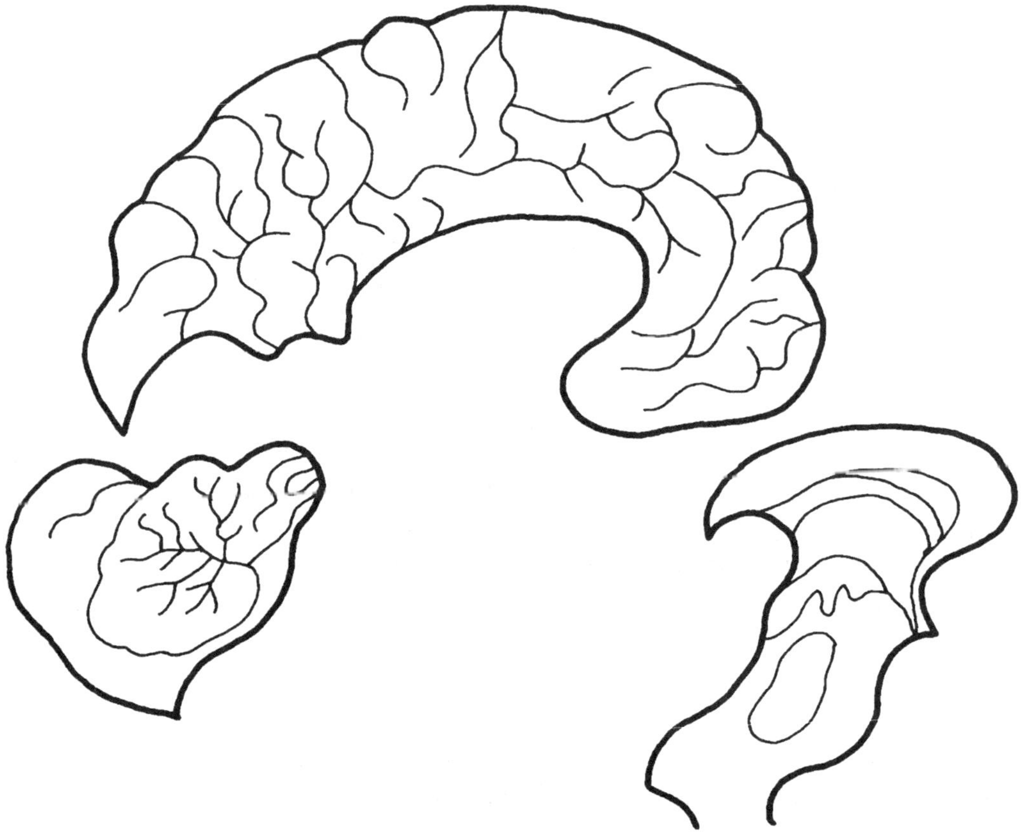

The Inside and Outside of Me

Science Concept: Besides the outside of my body that I can see, there is an inside part of my body that I cannot see.

Concept Literature: *What's Inside My Body* by Angela Royston and *Your Insides* by Joanna Cole

Question

Is there more to me than I can see?

Materials Needed for Each Child

- copy of *What Is Inside of Me?*, page 39
- pencil

Discovery Experience

This transition activity is designed to help your children make the switch from thinking about the characteristics of their outside bodies to wondering about what is on the inside of their bodies.

1. Designate one day as "inside-out day." Send a parent letter home a few days in advance so that your children may arrive at school on "inside-out day" wearing their clothes inside out.

2. On "inside-out day," gather your children around and enjoy the silliness of the event. Have one or two volunteers stand by you as you begin a discussion with questions like these: This is a silly day because we don't usually wear our clothes with the insides facing out. What can you see on (child's name) shirt that you can not usually see? (seams, button holes, tags, etc.) What can you see on his/her pants that you cannot usually see? Have you ever wondered what is on the inside of your body? Are there clues you can see or feel on the outside of your body that make you suspect there is an inside to your body too? (I can feel my bones. I had a cut and blood came out of my body. I can feel my heart beating when I run.)

3. Explain to your children that just like there is an inside part to our clothes that we do not usually see, there is an inside part to our bodies that we do not usually see.

4. On a teacher-made chart, ask your children to list parts that are inside their bodies. Save this chart and make another one just like it at the end of the body unit. Compare the two charts and use them as an assessment tool at the end of this unit.

5. After a thorough brainstorming session, give each child a pencil and copy of page 39. Using pencils, instead of crayons or markers, will encourage your children to draw details. Have children draw as many inside body parts as they can. Have them use arrows to label their drawings so the inside parts can be easily identified. Note: Save this page and repeat step 5 again at the end of the body unit. By comparing today's work and this same page completed at the end of the unit, you will be able to see how each child's ideas about the inside of the human body have changed and increased.

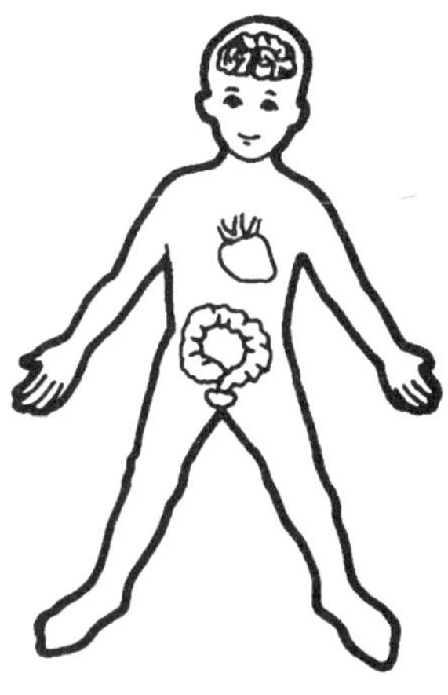

The Inside and Outside of Me *(cont.)*

What Scientists Know

Your body is an amazing machine. It all works together to keep you alive. You probably spend a lot of time thinking about the outside of your body: What should I wear today? I need a hair cut. I have a cut on my knee. But the inside of your body is just as important as the outside. Inside your body, there are systems that help you to do things like eat and breathe. If it were not for our insides, we would not have outsides.

Real Life Application

You can get a glimpse of the inside of the body by looking down through the throat. Give your children flashlights and ask them to peer into a partner's throat. Ask questions like these:

What color is the skin on the inside of the mouth?

How is the inside of your cheek different from the outside?

Describe what you see at the back of the throat.

Read *Your Insides* or *What's Inside My Body* to introduce the children to their insides and prepare them for the remainder of the lessons.

Complete page 1 of *The Inside and Outside of Me* Do-Along Book.

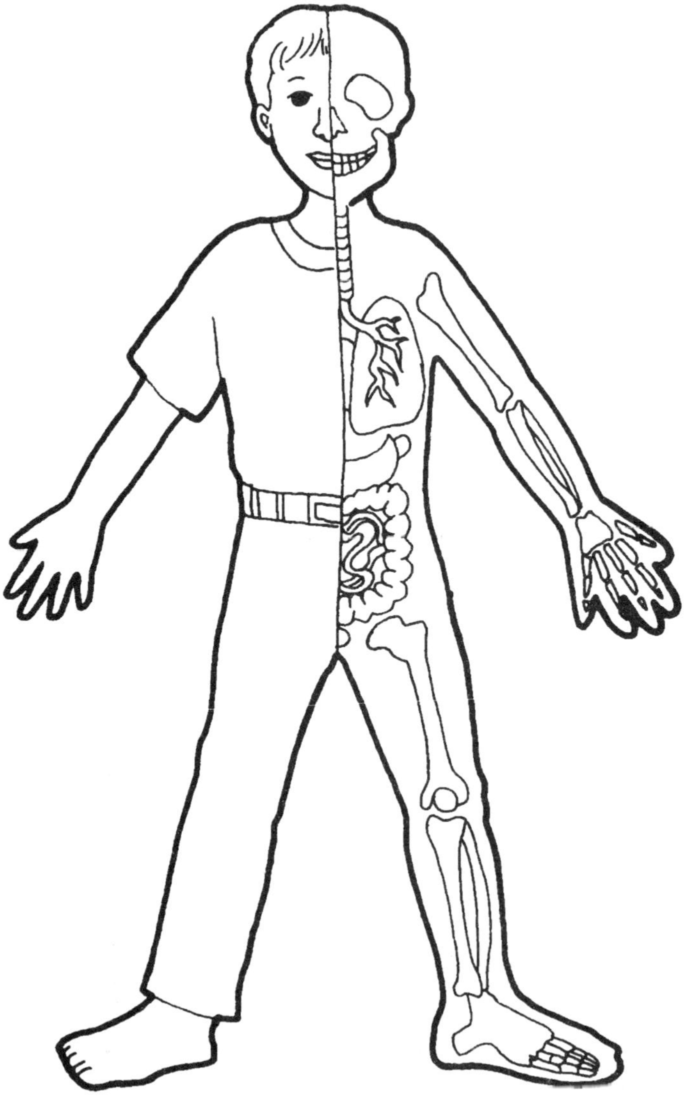

Body Builders

Show your children how their body outlines open up to reveal insides. If you have already constructed a model for display, open it to show your children what they will be making.

The Inside and Outside of Me *(cont.)*

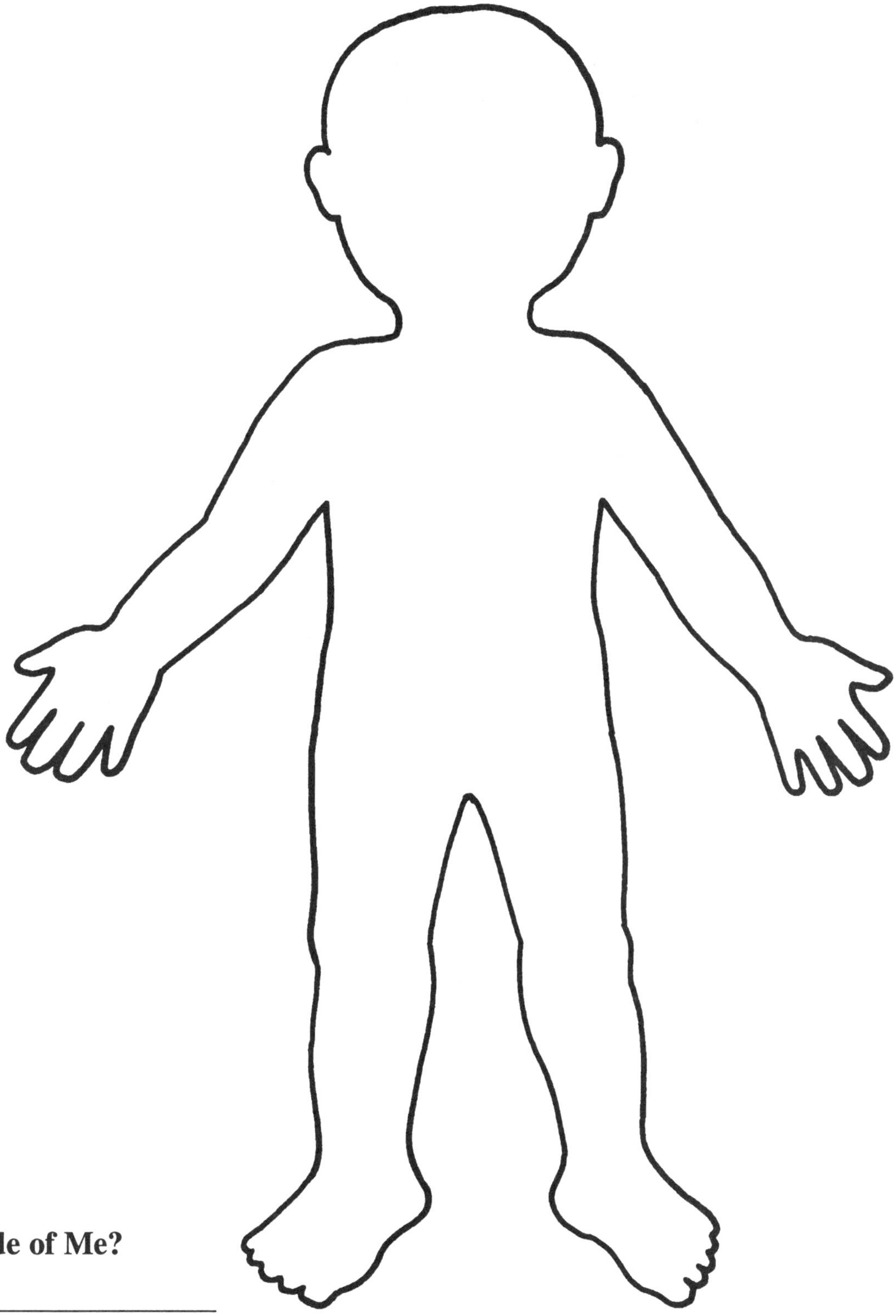

What Is Inside of Me?

☐ pre-test ____________________
(date)

☐ post-test ___________________
(date)

Every Body Has Five Senses

Science Concepts: I have five senses that help the inside me learn about what is happening with the outside me.

Concept Literature: *My Five Senses* by Aliki

Question

How do my five senses help me discover what is around me?

Materials Needed for the Class

- *My Five Senses* by Aliki
- a popcorn popper
- unpopped popcorn
- oil

Discovery Experience

While your children are out of the room, prepare a popcorn popper with oil and popcorn so all that needs to be done is to plug it in. Set the popper within hearing range of your story-circle area, but hide it behind a screen or shelf so that it is not visible to your children. Ask your assistant to discretely plug the popper in while you are talking to your children and reading *My Five Senses* by Aliki. You want the popcorn to begin to pop during the story.

1. Gather your children around. Remind them that their bodies have outsides and insides. How do you think the outside of you communicates with the inside of you? How do you think the inside of you knows what is happening on the outside? Well, today we will find out that the inside and outside can communicate. One way your body communicates is through the five senses. The five senses are like the windows of our bodies. Show your children the cover of *My Five Senses* and ask if anyone can name one of the five senses.

2. As you begin to read the story, the popcorn will begin to pop. At first your children will continue to sit quietly until they are sure something funny is happening. Continue reading until your children demand an explanation for the POP! POP! and the wonderful smell.

3. Stop reading and allow your children to discuss what they believe is happening behind the shelf or screen. How do you know there is popcorn popping in the room? Which of the five senses helped you to figure out there was popcorn popping in the room?

4. Bring out the popcorn popper. Now you know that it is popcorn because you can see it with your eyes.

5. What is your body telling you to do with this popcorn? Which of the five senses will you use as you eat and enjoy this popcorn?

6. Pass out the popcorn and let your children touch, taste, smell, and look at it as you finish reading the book.

Everybody Has Five Senses *(cont.)*

What Scientists Know

While I was reading the book, your ears heard a popping sound so your ears sent a sound message to your brain. Your brain wanted to know more about the popping sound, so it asked your five senses to find out more. After a few moments, your nose was able to smell the popcorn. So, quick as a wink, a smell message went up to your brain. You have probably eaten popcorn before so your brain was able to figure out that popcorn was being made somewhere in the room. Once you knew it was popcorn, you began to raise your hand and whisper to your neighbor. At this point your brain began to prepare your stomach and mouth for a snack, "Hey, stomach," your brain said, "get ready for some popcorn."

While you were eating the popcorn, you were touching it, looking at it, smelling it, tasting it, and listening to the crunch. Your brain took in all these signals, too. Your five senses are like detectives for the brain; they gather information and clues to help the brain put it all together for you. So, if you knew that popcorn was popping in the room, thank your five senses and your brain. **Note**: Although the brain has not been introduced, most children are aware that they have a brain and that it helps them to think. The next activity will elaborate on the brain and its function as "the boss of the body."

Real Life Application

Help your children tune into their five senses. Pass around a cozy stuffed animal. Ask them which of their five senses they are using to enjoy and gather information about this item. Pass other items around and talk about exploring each one with the five senses. Complete page 3 of *The Inside and Outside of Me* Do-Along Book to make a scratch and sniff sticker.

Brain in Charge

Science Concept: The brain tells every part of my body what to do.

Concept Literature: *Thinking* by Kathie Billingslea Smith

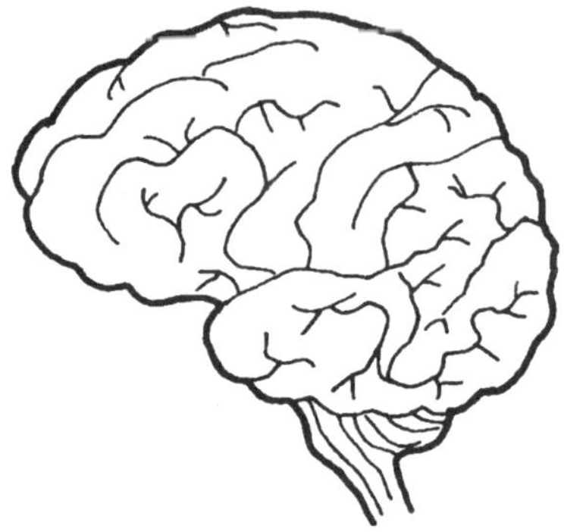

Question

Why is my brain important?

Discovery Experience

This game, a variation of Simon Says, will help your children to understand that the brain is the boss of the body.

1. Ask each child to point to his/her brain. Tell your children that their brains are hidden under a thick bone called the skull. Explain that the brain is the boss of the body because they need their brains to think, talk, read, run, dream, love, color, cry, breathe. . . They need their brains for everything they do.
2. Gather your children in a circle to play "The Brain Says." To play, explain that every command that is preceded by the phrase "the brain says" should be followed. Commands that are not preceded by this phrase should be ignored. Remember, if the brain does not tell you to do it, you cannot do it. A sample round may go something like this:

The brain says, "Scratch your head." The brain says, "Hop on one foot." The brain says, "Think of something good to eat." The brain says, "Say hi to your neighbor." The brain says, "Laugh." Raise your right hand. Whoops! If you raised your right hand, you should not have. Remember, the brain is the boss of the body and you cannot do anything until the brain tells you to.

What Scientists Know

Scientists have been using their brains to think about their brains for a long time. But, there is still much about the brain that scientists do not know. The brain is very complicated, but scientists do know that your brain is busy working all the time, storing information and directing your body. Some people say the brain is like a computer, but it is much more powerful than the best computer in the world. Your brain is what makes you special and unique. Your brain is a very important part of your body.

Real Life Application

The brain has many functions. Memory is but one of its functions. Cut out different body part pictures (two of each part) from magazines and glue them on index cards. Play concentration with the body cards. Place all the cards, pictures down, on a table. Children flip two cards over, hoping to get a match. When they do match two cards, they remove those cards from the table and put them in a pile to keep track of their scores. Give your brain the credit every time you remember and are able to make a match.

Complete page 3 of *The Inside and Outside of Me* Do-Along Book.

Body Builders

Add the brain to your body. See card 4 for instructions. Show your children how the brain takes in information.

Brain Puzzle

Science Concept: The brain has three main parts with different jobs.

Concept Literature: *How the Body Works* by Steve Parker

Questions

How do the parts of the brain fit together? What are their individual jobs?

Materials Needed for Each Child

- *The Inside and Outside of Me* Do-Along Book
- puzzle pieces, page 36
- scissors
- glue

Discovery Experience

1. Ask your children if they enjoy putting puzzles together.
   Explain that they will be putting together a puzzle of their brains.
2. Instruct your children to cut out the three puzzle pieces and turn to page 4 in their do-along books.
3. Lead children through this activity by using this dialogue:

Pick up the big piece with the wrinkles on it. What you are holding in your hand is a picture of the cerebrum, the biggest part of your brain. This is the thinking part of your brain. Can you match its shape to a shape on the puzzle? Glue your cerebrum in its place. Make sure children place it in the correct location. Now, pick up the smaller round piece. Meet the cerebellum. The cerebellum helps you balance and keep your posture. It also helps you move in fancy ways. The cerebellum is located in the back of your brain. Can you find a matching round space on your puzzle? Glue it in place. Look at that last piece. It is long and thin. This is a picture of the brain stem. Your brain stem is very important. It is the part that connects right to the spinal cord. The brain stem controls body functions like breathing and keeping your heart beating. Glue the brain stem onto your brain. Your brain is amazing.

What Scientists Know

Scientists know that they do not know everything about the brain. They have been able to divide the brain into the three main parts which you have glued onto your puzzle. This puzzle does not show the brain exactly as it appears in your body. It shows what the brain would look like if it was sliced in half and opened up so that you could see inside. If you saw a picture of the outside of your brain, you would see almost nothing but the cerebrum. The cerebrum is so big, it overlaps to cover almost all of the other two parts. If the cerebrum was unwrinkled and unfolded, its surface would be about the size of the top of your teacher's desk or a desk in an office.

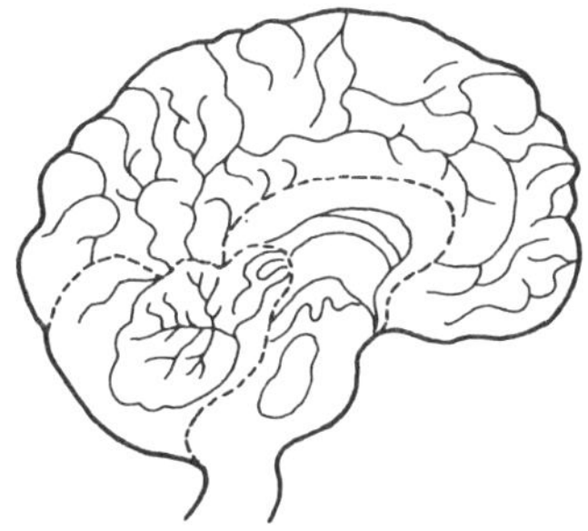

Real Life Application

Use this dialogue to review the function of each main brain part. Instruct children to point to the part on their puzzles that answer the following questions: Think about a friend. What part of your brain did that? You are breathing. Which part is at work? Balance on one foot. Which part of your brain is in control?

Send a Message

Science Concept: Nerves carry messages between the body and the brain.

Concept Literature: *The Magic School Bus Inside the Human Body* by Joanna Cole

Question

How does my brain know if the water I touch with my finger is warm or cold?

Materials Needed for Each Group

- container of warm water
- container of cold water
- The *Inside and Outside of Me* Do-Along Book, page 5

Discovery Experience

1. Divide your class into small groups. Give each group one container of cold water and one container of warm water. Ask your children if they can tell by looking, which container has the warm water and which container has the cold water.
2. Pose a silly question to introduce the nervous system. If your brain is the part of your body that decides which container has the hot and cold water, should you stick your brain (head) in the water so that it can decide which is the warm water and which is the cold water?
3. Ask your children to explain an easier way to test the water temperature. Encourage them to try to explain how the brain can know the temperature of the water if it is the finger that will actually be in the water.
4. Encourage each child to test the water by sticking one finger in each container. Allow each group to report its findings.

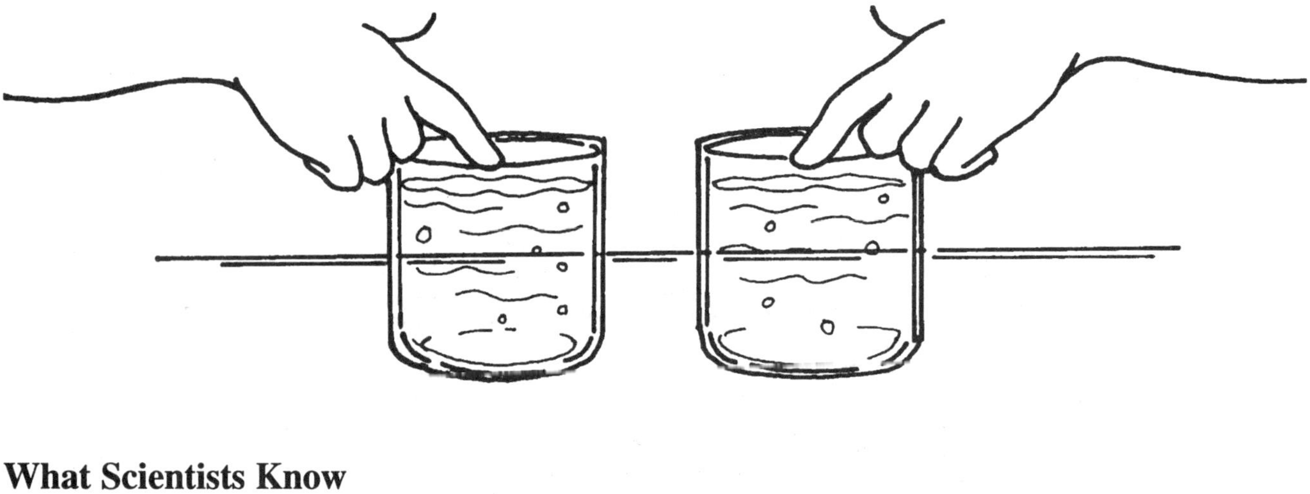

What Scientists Know

Your brain and your finger each have an important job to do today. Your brain made the decision, but your finger did the water test. What makes the connection between the finger and the brain? Inside your finger there are nerves that run from your finger and up your arm to your brain. Just like the mail carrier or telephone wires, nerves carry messages from the finger to the brain. Nerves, however, are much faster than mail carriers or telephone wires. You did not need to get your head wet for your brain to quickly know which water was warm and which was cold.

Send a Message *(cont.)*

Real Life Application

How quickly can a message get from your brain to your hands? Play "Slap It" to find out. Find a partner and stand face to face. Hold out your hands, palms down. Your partner must try to slap the back of your hands before you move them out of the way. What is happening in this game? Your eyes sense when your partner is about to slap your hands. So, they flash a signal to your brain, "Here comes a slap!" When your brain becomes aware of the situation, it sends a message to the muscles in your hands. Flash! You are able to move your hands out of the way.

Complete page 5 of *The Inside and Outside of Me* Do-Along Book.

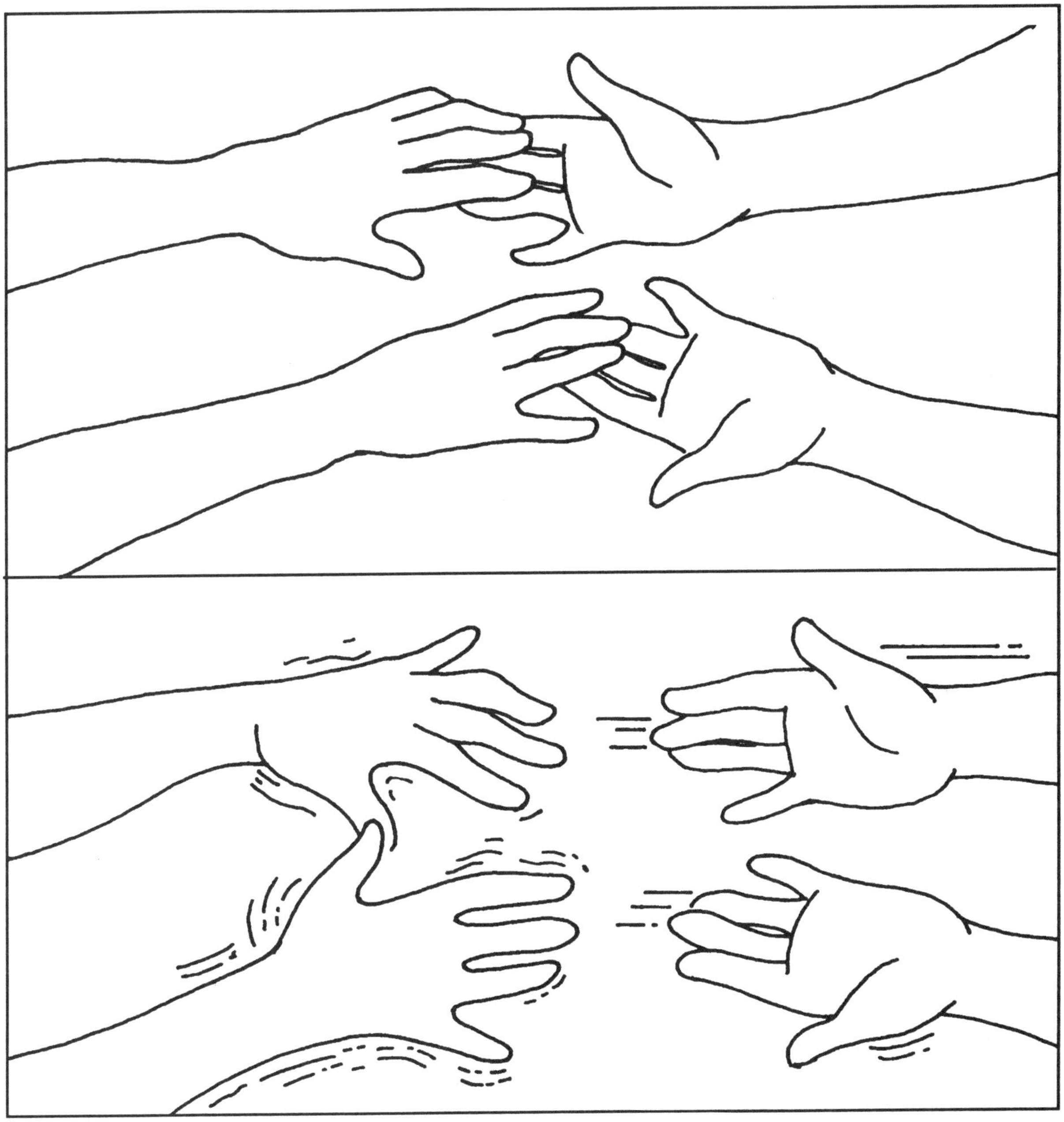

Nerves and Telephone Wires

Science Concept: The nerves carry impulses to and from the brain. The spinal cord is a thick bundle of nerves that runs up your back.

Concept Literature: *Touching a Nerve: How you Touch, Sense & Feel* by Steve Parker

Question

What is inside the bumpy bone on my back?

Materials Needed for Every Two Children

- two plastic cups
- craft knife
- a 1.09 yard (1 m) piece of string
- two toothpicks

Discovery Experience

Note: To make a hole in the bottom of a plastic cup, poke a craft knife through the bottom of the cup and make a small hole.

Begin today's lesson by telling your children there is a telephone system inside their bodies. After introducing the idea that there is a telephone system (the nervous system) inside the body that the brain uses to communicate, follow the directions to make cup and string telephones.

How to Make a Telephone

1. Insert a string into the hole at the bottom of each cup. Tie a toothpick onto the string in each cup so the knots will not slip out of the cups.

2. A voice will only carry on a taut string, so pull on the strings while talking and listening.
3. Allow your children plenty of time to explore with their cup telephones. How does your voice get from one cup to the other? Help your children to understand how the nervous system works by comparing it to telephone wires. Tell your children that inside your body there are many nerves. Nerves are like telephone wires that your brain uses to communicate with the rest of the body.
4. Show each child how to locate his/her partner's backbone. Encourage your children to feel the bumps of the backbone that run from the waist all the way up to the brain.

Nerves and Telephone Wires *(cont.)*

What Scientists Know

Sound travels along the string from one cup to another as your voice causes the string to vibrate. The spinal cord carries impulses to and from the brain. The spinal cord is kept safe inside the bumpy bones that you can feel on your partner's back. The spinal cord is like the string between the cups. It carries messages between your body and your brain.

Real Life Application

Send a squeeze! Have your children sit in a circle for this simple game to show how messages travel. Join your children in the circle and begin by squeezing the hand of the child next to you. He or she should squeeze the hand of the next child and so on until the "squeeze" comes back to you. Start out slowly and then try to send a squeeze very quickly. Explain that messages in the body travel much faster than we could send a squeeze. Read pages 5, 6, and 7 of *The Inside and Outside of Me* Do-Along Book to reinforce that nerves carry messages.

Body Builders

Add the spinal cord and nerves to your body outline. See card 5 for directions.

Just the Facts

This section of *Our Bodies* uses the five senses as windows into the internal workings of the body. Before each system is introduced, your children will have an opportunity to explore one of the five senses that relates to the body system. For example, the sense of smell is related to the respiratory system which in turn is connected to the circulatory system. The purpose of this format is to help children make connections between that which is familiar to them (the five senses) and that which is unfamiliar (their inner workings). The format of these lessons provides a variety of ways your class can gain an understanding of their bodies. Each time you follow the directions on the Body Builders Cards to add a system to the life-size working model, an activity is included designed to give your children hands-on experience and a visual representation of their internal systems.

Teacher Information:

Note: You will find information about the body written at a child's level after each activity.

Traditionally, the five senses include sight, hearing, smell, taste, and touch. Each of the sense organs receives information, converts it into nerve signals, and sends it to the brain for interpretation. To avoid information overload, the brain discards most of the information it receives from the five senses. You are made aware of only the information the brain deems important. There are other sense organs in the body as well. For example, there are stretch sensors in muscles that tell the brain about the position of the body.

The body is made up of systems. Systems are made up of organs. Organs are composed of tissues. The cell is the basic building block of tissues. Cells contain organelles, and each one contains many molecules. Molecules are a combination of atoms. Therefore, bodies are made up of tiny particles called atoms.

Each of the body's systems has a job to do. The systems included in *Our Bodies* are:

- **nervous system**—the brain, spinal cords, and nerves
- **skeletal system**—the bones which support
- **integumentary system**—the skin which protects and contains
- **muscular system**—the muscles which allow movement
- **respiratory system**—brings oxygen into the body
- **circulatory system**—delivers oxygen and nutrients
- **digestive system**—prepares food for the body

The urinary, reproductive, hormonal, and lymphatic systems are outside the scope of this book.

Body systems work together to keep the body alive and functioning.

My Five Senses Tell Me About the World

(For general book-making instructions, see page 9.)

Supplies needed for this do-along book:

- magazines
- aluminum foil
- glue
- scissors
- spray air freshener
- stapler
- thin rubber bands
- chewing gum
- sandpaper
- silky fabric
- corduroy fabric
- crayons
- pencils

Cover: Finding body parts in a magazine.

Children may cut out pictures from magazines of the body parts that help us to do each of these five actions: seeing, hearing, smelling, tasting, and touching. They should glue the pictures in the appropriate boxes on the cover. Or, instead of cutting out pictures, the five body parts may be drawn. Children should write their names at the bottom of the page.

Page 1: Look in a mirror.

Read the page with your children. Cut out an aluminum foil rectangle to glue over the rectangle on page 1. Encourage children to use their eyes to see inside.

Page 2: Listen to a vibration.

Read the page with your children. Staple one thin rubber band at both of the points on page 2. Encourage children to listen with their ears as they gently "twang" the rubber band.

Page 3: Smell a scent.

Read the page with your children. Cut out a piece of paper the same size as the box on page 3. Spray a bit of air freshener or perfume on the paper and ask children to glue it into their books over the box. Have children smell the box.

Page 4: Taste a raisin or cracker.

Read the page with your children. Ask your children if they like raisins and crackers. Offer raisins or crackers to your children, pass out the raisins and crackers, and save the wrappers. As they are eating their snacks, instruct children to glue their wrappers into the mouth on page 4.

Remind them that we really never eat wrappers, just the raisins and crackers. Ask them if they like the taste of their snacks.

Page 5: Feel a texture.

Read the page with your children. Glue a piece of sandpaper, silky fabric, and corduroy (these are just suggestions—use three different items with texture) to the boxes on page 5. Encourage your children to feel the different textures with their fingers.

My Five Senses Tell Me About the World (cont.)

1

I see with my eyes.

Glue Aluminum Foil Here

LOOK AT ME!

MY FIVE SENSES TELL ME ABOUT THE WORLD!

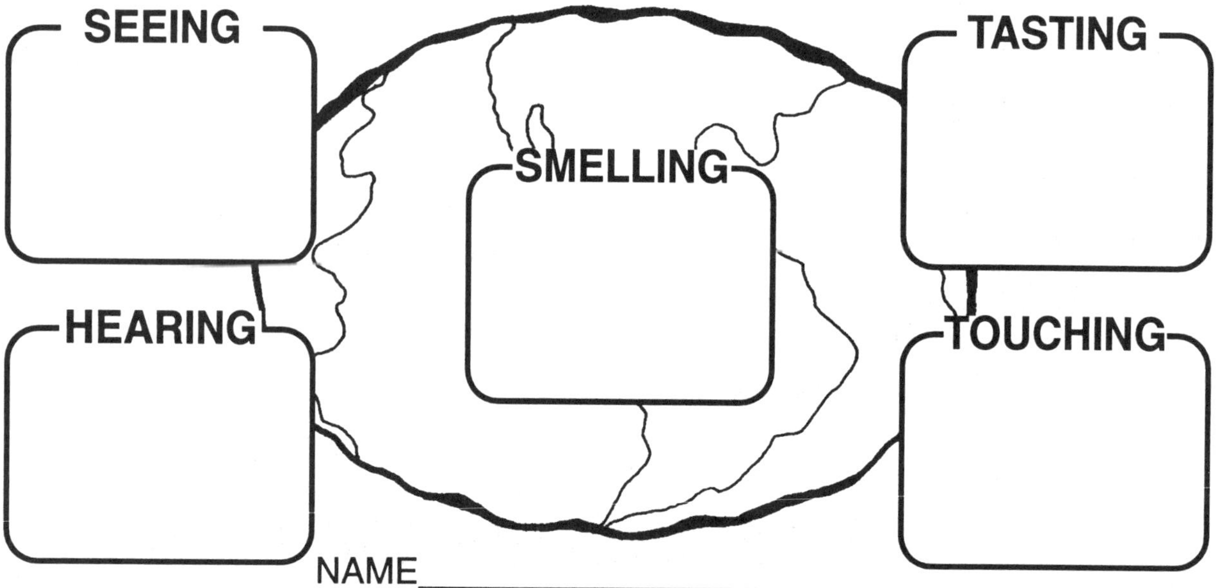

NAME________________________________

My Five Senses Tell Me About the World *(cont.)*

3 **I smell with my nose.**

LISTEN TO THIS!

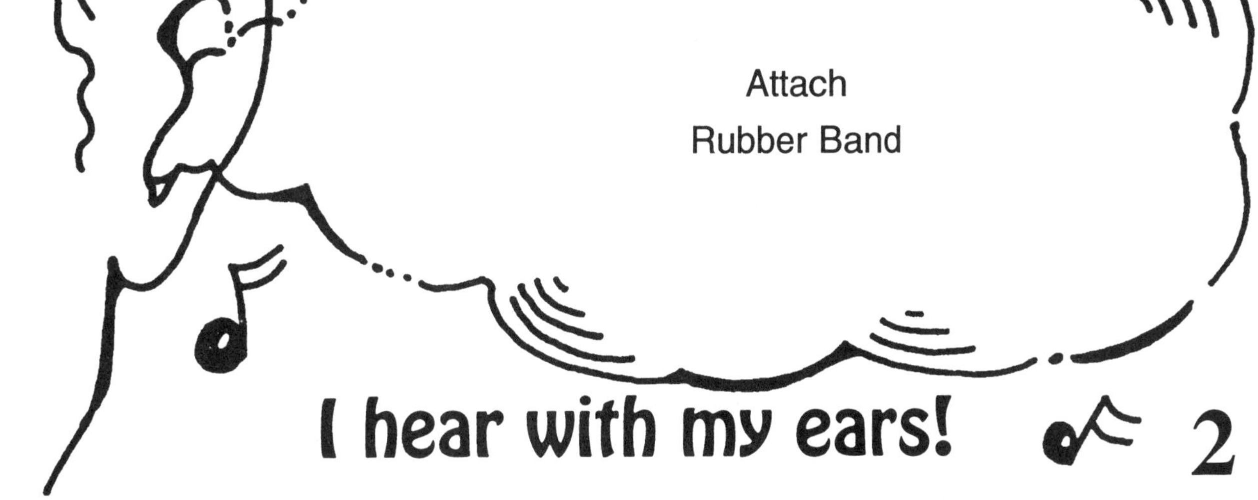

I hear with my ears! 2

My Five Senses Tell Me About the World (cont.)

I touch with my fingers. 5

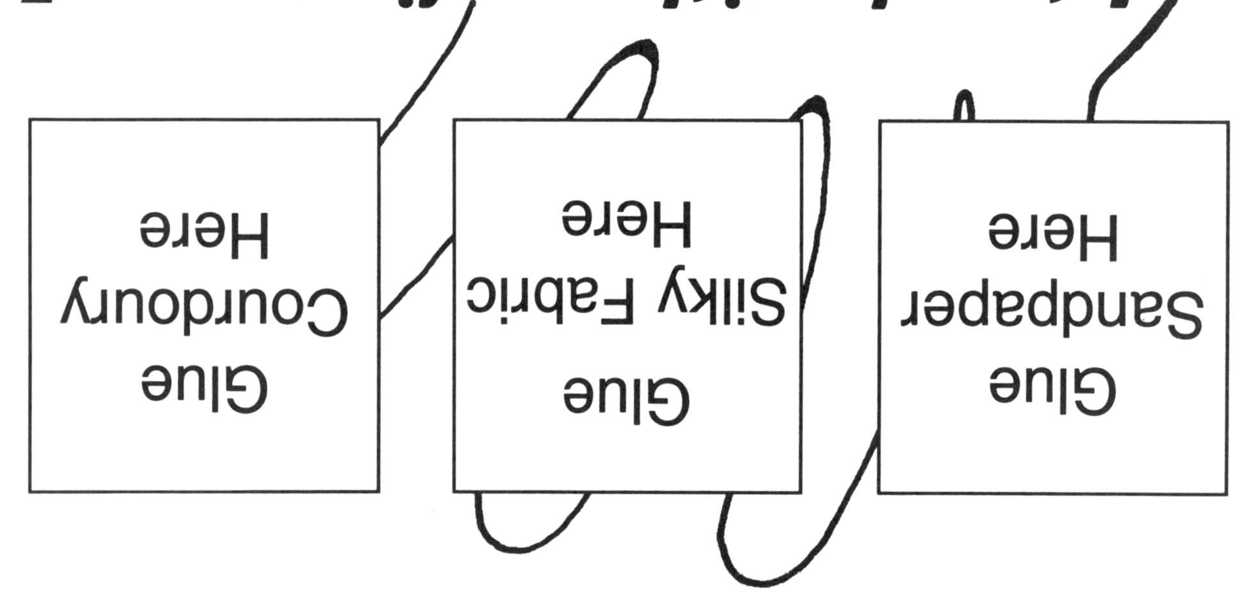

YUM! YUM!

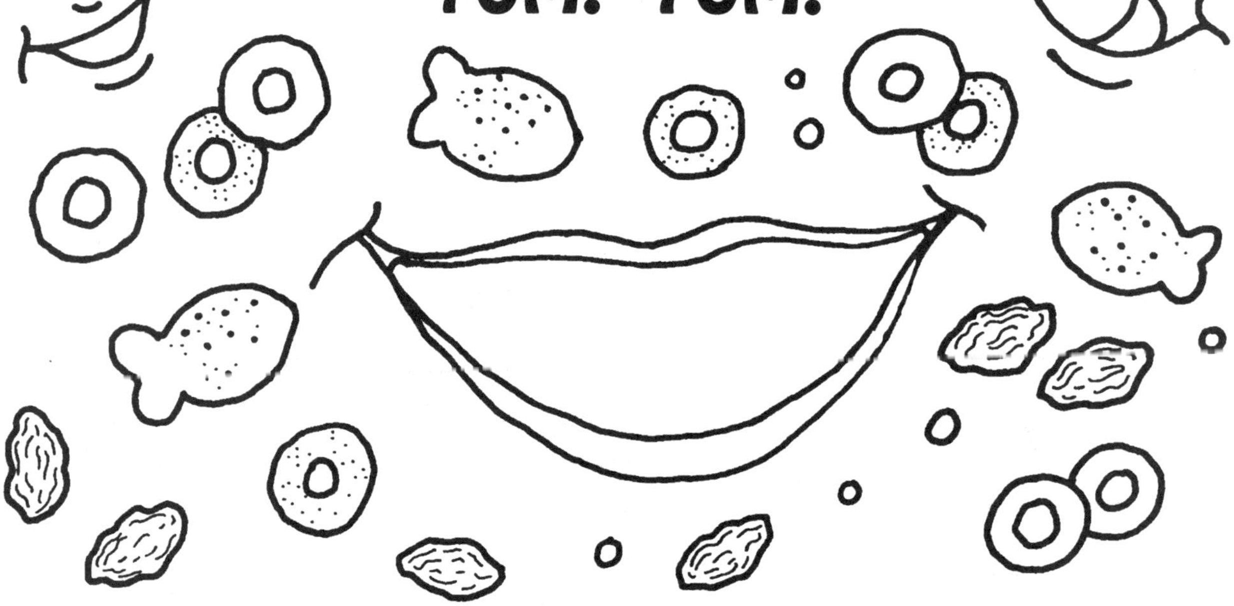

I taste with my tongue. 4

I Can Tell by Touching

Science concept: My sense of touch tells me about the world.

Concept Literature: *The Touch Book* by Jane Belk Moncure and *Find Out by Touching* by Paul Showers

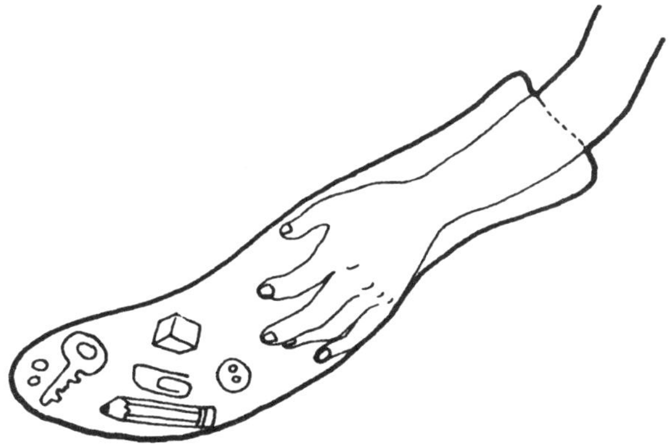

Question

Can I identify objects using only my sense of touch?

Materials Needed for the Class

- a copy of page 55, cut into cards
- a feather

Materials Needed for Each Group

- a tube sock
- choose several of the following items: bolt, rock, block, button, paper clip, key, cotton ball, wad of paper, unsharpened pencil, bean or seed, crayon, rubber band, craft stick, bead, piece of tape, eraser

Discovery Experience

Prepare for this activity by filling tube socks with some or all of the materials listed above. Cut apart the cards on page 55 that represent each item you have placed in the sock. (Note: If you have a large class you may choose to make one touching sock for every two children and have them share.)

1. Gather your children into a circle. Walk among your children and tickle each one with the tip of a feather. Tickle some in very sensitive spots like the neck, cheek, or nose. Tickle some in less sensitive spots like the shoulder or back. As you tickle each child ask, "Does that make you smile?" or, "Can you feel my tickle?" Help your children to conclude that their skin is feeling the tickle, but some locations are more sensitive than others.

2. Explain to your children they were able to feel the tip of the feather because of tiny feeling points or sensors that lie just under the top layer of skin. (The outer layer of skin is called the epidermis, and the inner layer is called the dermis.)

3. Have each child lightly touch his/her finger tips with his/her thumb. The fingertips are very sensitive. Why do you think fingertips need to be sensitive? "Let's try an activity to help us think about this question."

4. Give each child a tube sock full of some or all of the items listed above. Sit in a circle. Allow your children to feel inside the sock without looking or pulling anything out. "Can you guess the identity of some of the items in the sock?"

5. Show your children one of the item cards from page 55. Challenge them to pull the specified item out of the sock, using only their fingertips to identify the object. Continue the game until all the items have been removed.

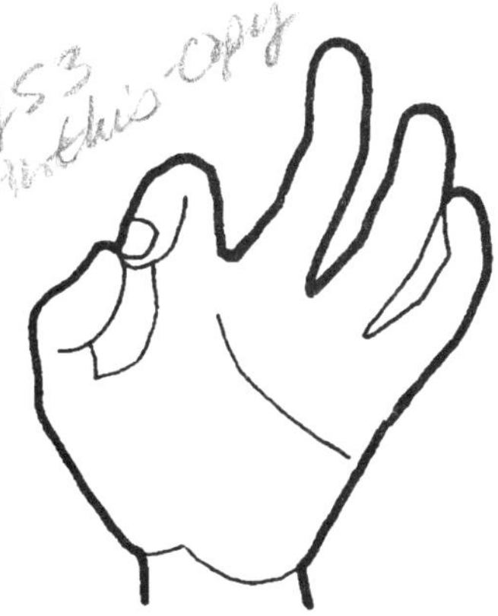

I Can Tell by Touching *(cont.)*

What Scientists Know

Your skin has many feeling points. These points can feel a gentle squeeze, a painful pinch, or sensations of hot and cold. In your fingertips the feeling points are very close together, but on your back the feeling points are farther apart.

That is why some parts of your body are more sensitive to a feather's touch than others.

Real Life Application

Braille is a reading system for people who are blind. Patterns of raised dots are felt by the fingers. If possible, bring in an example of a Braille book for your children to read. Some libraries have Braille books or examples of Braille. Try this modified raised letter approach to allow your children to read their names with their fingers. Write each child's name on a large index card. Have each child trace over the letters in his/her name with white glue. For best results, keep the glue in a bead rather than spreading it out. Younger children will need assistance with this. After the glue has dried, your children can close their eyes and feel the glue in order to read their names.

Give each child an opportunity to select his/her name from a group of names without using his/her eyes.

Complete page 5 of the *My Five Senses Tell Me About the World* Do-Along Book.

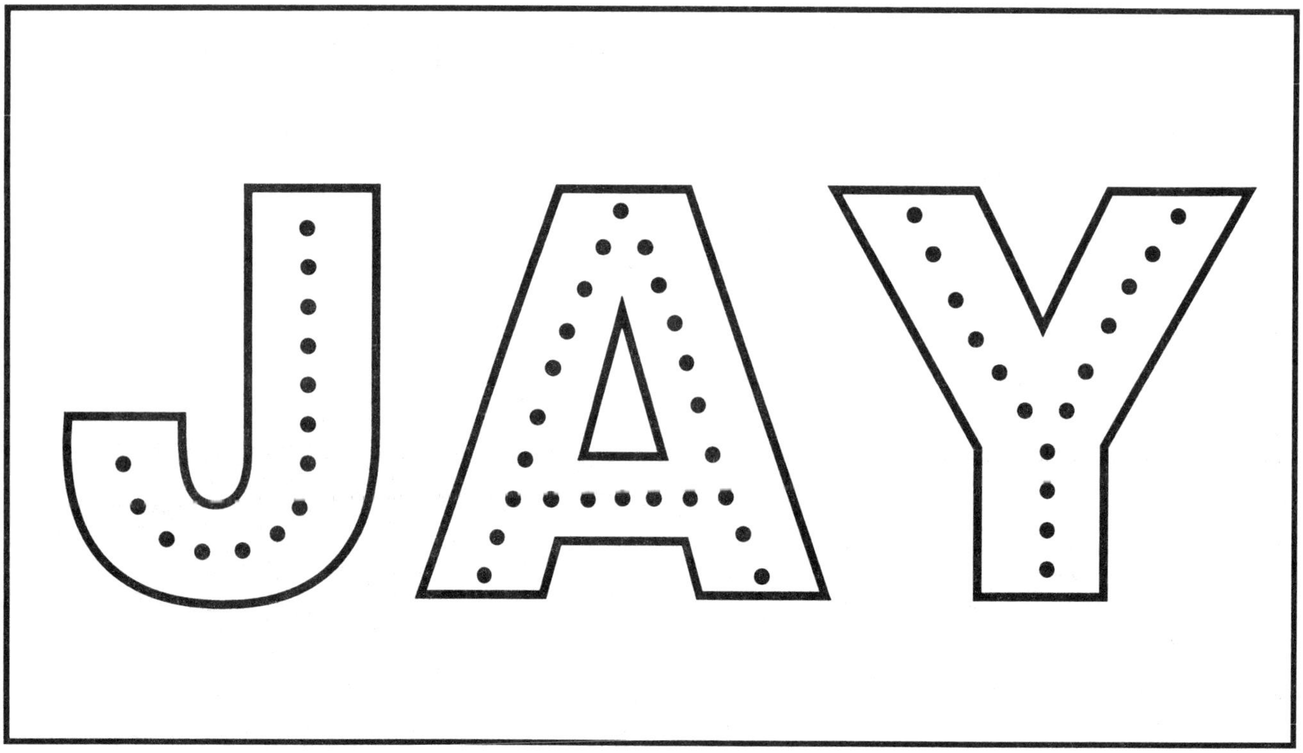

I Can Tell by Touching *(cont.)*

Directions: Cut out each card that matches an item that has been placed in a tube sock.

 craft stick	 button	 paper clip
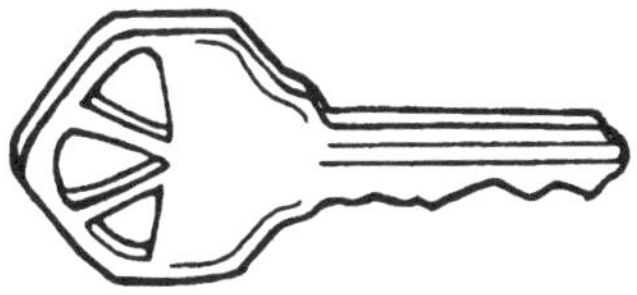 key	 screw	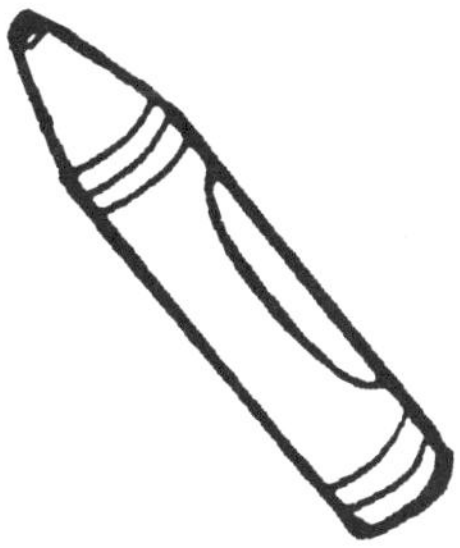 crayon
 wad of paper	 rubber band	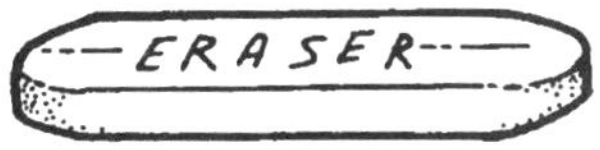 eraser
 seed	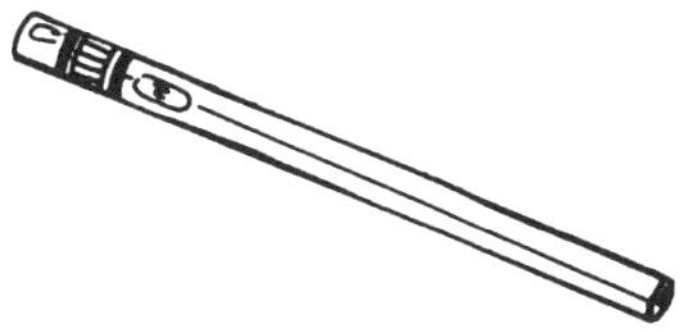 unsharpened pencil	 cotton ball

Skin—Up Close and Personal

Science Concept: Areas of skin on my body are customized to perform different jobs.

Concept Literature: *My Hand* by Aliki

Question

How is the skin on the back of my hand different from the skin on the palm of my hand?

Materials Needed for Each Child

- a magnifying lens
- a copy of page 58
- pencil

Materials Needed for the Class

- chart paper
- a marker

Discovery Experience

What I Know About My Skin
• On My Body • Get's Cut • Is Grown • Is Tan • Feels Soft

1. "My skin is a very important part of my body, but I don't think about it very much. Today, let's think about our skin." As your children list everything they know about skin, record their ideas on the top of a chart page entitled "What I Know About Skin." Introduce the terms palm of my hand and back of my hand so everyone is prepared to do this activity.
2. Pass out a magnifying lens and a copy of page 58 to each child.
3. Have everyone observe the back of his/her hand through the lens.
4. Provide pencils so that your children can draw the details of the backs of their hands on page 58. Using pencils instead of crayons will encourage your children to sketch details rather than color the picture.
5. Have everyone flip his/her hand over and study his/her palm.
6. Tell your children to record on the lab sheet what they see on their palm drawings.
7. Use these questions to help your children verbalize their findings: Where is your skin the smoothest? Why are there wrinkles on the palm of your hand? Is there hair growing on your hands? Where? Does one side of your hand have thicker, tougher skin than the other side? Why? Describe what you see on your fingertips. Describe any other interesting things you see while looking at your hands.
8. Encourage your children to use a magnifying lens to look at the skin on their arms, legs, etc. Not all skin is the same. Allow time for your children to add to the "What I Know About Skin" chart.

Skin—Up Close and Personal

(cont.)

What Scientists Know

Skin on the palm of your hand is different from skin on the back of your hand. The skin on the palm of your hand is thick and smooth because this area of your body is designed for hard work. Do you know someone who plays the guitar or another stringed instrument? You may notice tough, thick spots on their fingertips in the places they push down on the strings of their instruments. These thick spots are called callouses. Callouses are just one way the skin responds to hard work. Your skin has wrinkles and folds, too. Maybe you saw these wrinkles on your knuckles. The extra skin on your knuckles allows your fingers to bend comfortably. Can you find another place on your body with this wrinkled extra skin?

Real Life Application

Using the following method to make fingerprints will allow your children to see the intricate patterns on their fingers. First, make a spot of lead on a piece of paper by rubbing the side of a pencil on the paper. Rub one child's index finger in the lead until the fingertip is dark and well-covered. Transfer the print to a piece of tape by placing the index finger onto the sticky side of the tape. Remove the tape from the finger and stick it on a piece of white paper. The fingerprint will show up underneath the piece of tape. Use a magnifying lens to appreciate the uniqueness of each print.

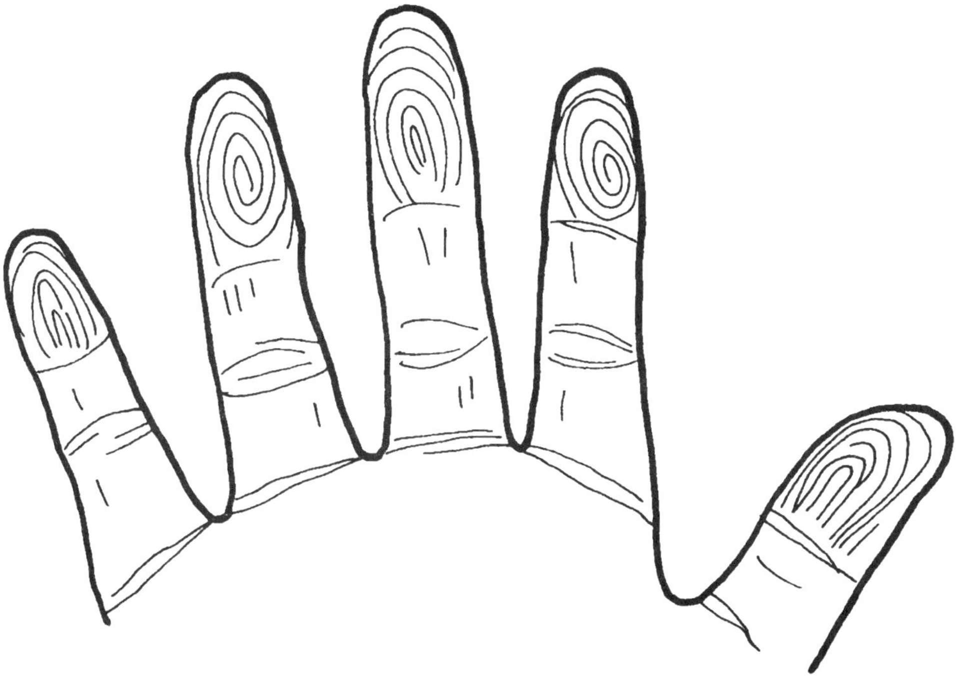

Invite a musician into your classroom. Ask him/her to show your children how quickly his/her hands move over the strings of the instrument as he/she plays. If the guest is willing, allow your children to press down on the strings of the instrument. Ask your children if their fingers hurt. Have someone ask, whether the guest's fingers hurt when he/she first began to play the instrument. Does it hurt him/her to play now? Have the musician play a piece of music and celebrate all that skin and hands can do.

Skin—Up Close and Personal *(cont.)*

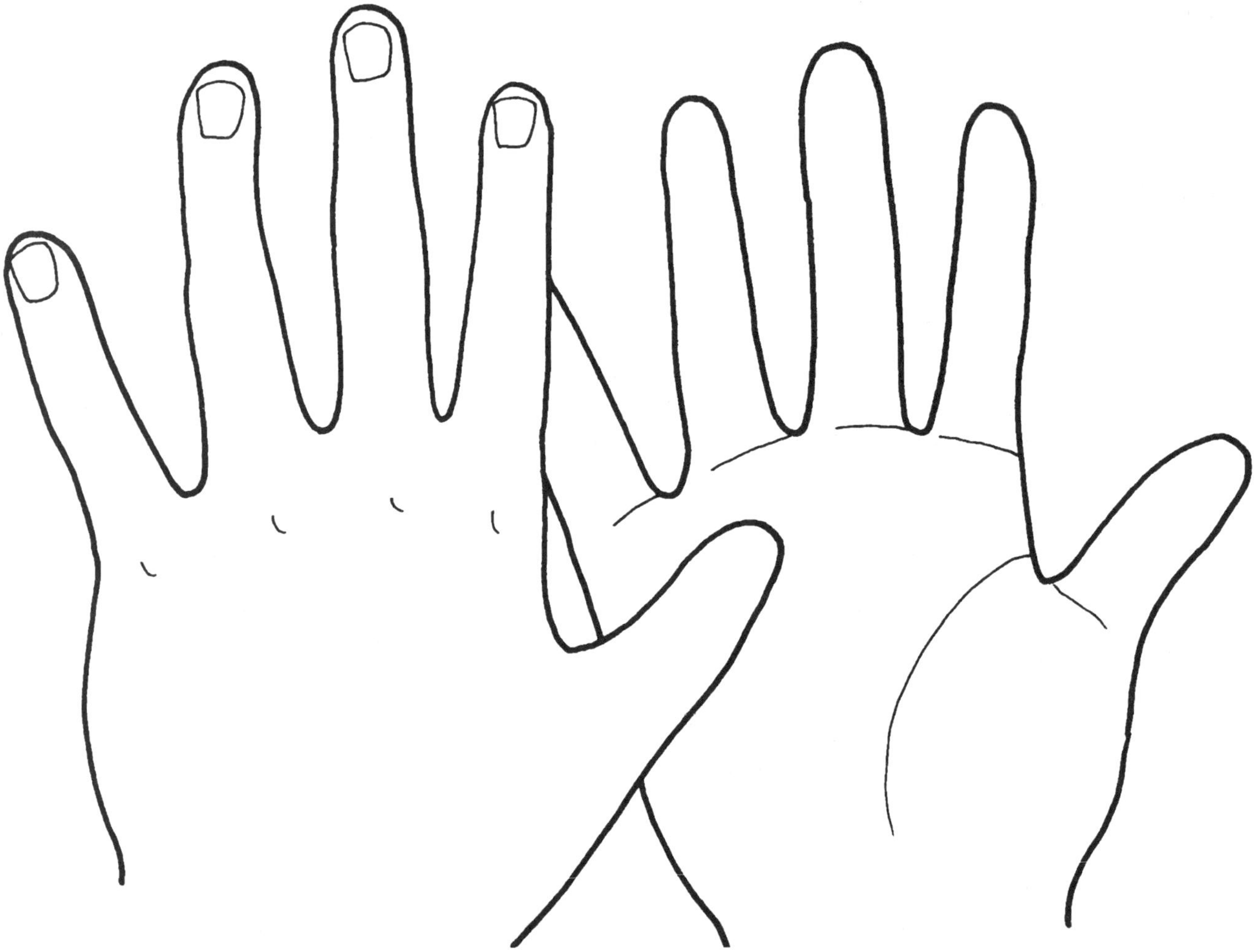

1. at your
2. what you see
3. at your
4. what you see

Skin Tents

Science concept: My skin is supported by my bones.

Concept Literature: *The Skeleton Inside* You by Philip Balestrino

Question

Why is my skeleton important?

Materials Needed for the Class

- old sheet or blanket
- a hanger

Discovery Experience

1. Clear a large area in your classroom of all furniture and clutter. A large story rug works nicely. Select a group of three or four children to be builders and ask your other children to sit on the edge of the rug. Provide the builders with an old sheet or blanket.
2. Challenge the builders to build a tent with the sheet. There are two rules for this building project. First, the group must stay within the designated boundaries. They can use only the materials you supply. Second, when the tent is finished, all the builders must be able to sit under the tent with their hands folded and without the tent touching their heads. After the first building team has tried, switch building teams and allow everyone a chance to build a tent.
3. Gather everyone around and allow each to share his/her frustrations. What would make this job easier?
4. Repeat the activity but provide the builders with supports such as chairs, chart stands, etc., to use as they build. Was it easier to build a tent this time? Why?
5. Present your sample body outline or one of your children's body outlines that was made in the first section, "Why Is My Body Special?"
6. Hold it above your head and ask "What will happen if I let go of (child's name)?" (Use the name of the child whose body outline you are holding.)
7. Drop the body outline and watch as it crumples on the floor.
8. Explain to your children that this is exactly what would happen to you if it were not for the bones that hold up your body. Tie this thought into the tent building experience.
9. Tape the shoulders of the body outline temporarily onto a hanger and hold it up for everyone to see. Tell your children that bones are like a hanger. They hold up the body.

What Scientists Know

Under your skin and inside your body are many bones (206). Bones hold up your skin and give your whole body its shape. Without bones you would crumple up, just like the body outline.

Real Life Application

Bones are throughout your body. Have your children feel and squeeze several different parts of their bodies for bones. Where is the one place on your body that you cannot feel any bones?

Dem Bones

Science Concept: Bones protect me, support me, and help me move.

Concept Literature: *The Bones Book* and *Skeleton* by Stephen Cumbaa and *Skeletons! Skeletons!* by Katy Hall

Question

What do bones do in my body?

Materials Needed for Each Child

- a copy of page 62

Discovery Experience

Set up these science stations and host a "Bone Day" in your classroom. Have your children visit each station to learn about bones and the skeletal system. As they circulate, they should record their ideas on page 62. "Bone Day" is a great day to invite parents to come in and help. Parent volunteers can sit at each station.

If a station format does not fit into your schedule, the following lessons can each be adapted and taught as regular lessons or circle time activities.

Station One

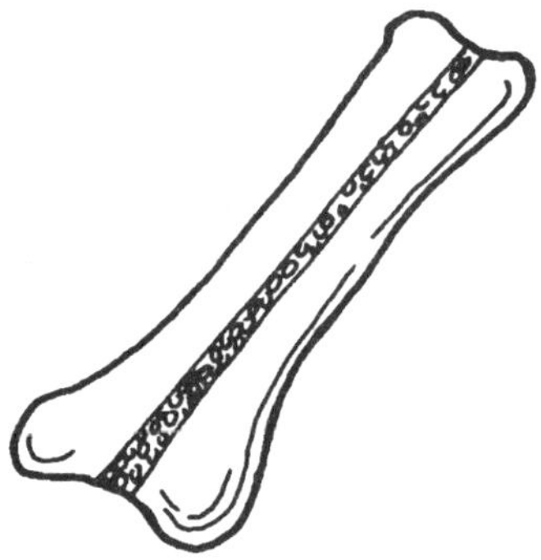

Materials Needed

- a soup bone
- magnifying lens
- colored pencils

1. Obtain a large soup bone from a butcher.
2. Ask the butcher to cut it in half the long way.
3. Set out the bone and a hand lens at this station.
4. Touch the outside of the bone. How would you describe it? Color a picture of what is inside the bone on your lab sheet.

Station Two

Materials Needed

- a realistic picture of a human skeleton (*Your Insides* by Joanna Cole has a simple skeleton drawing)
- arm and leg bones, pages 63 and 64
- Body Builder Card 3

Follow the directions on Body Builder Card 3 to glue paper bones to each child's body outline. They will have two arm bones on each arm and two leg bones on each leg.

Station Three

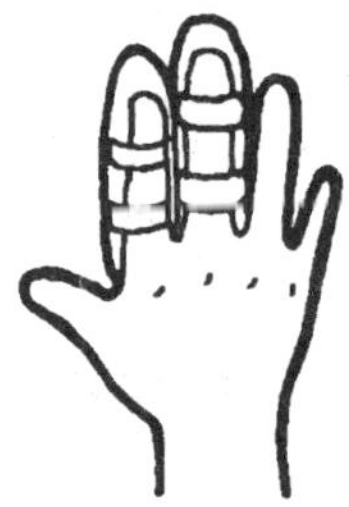

Materials Needed

- a pencil
- craft stick
- masking tape

1. Start out by having each child bend over and touch his/her toes and then twist at the waist. Bones cannot bend so you have joints to help you turn and twist. Your hands have many joints.
2. Show each child how his/her fingers bend around a pencil. We can twist, write, and catch a ball because our skeleton is jointed.
3. Now, take away the child's joints by taping a craft stick to at least one of the child's fingers.
4. Then, set out a pencil and let him/her try to write. Help each child to verbalize how difficult it is to write without being able to bend his/her fingers. Complete the lab sheet for this station.

Dem Bones *(cont.)*

Discovery Experience *(cont.)*

Station Four

Materials Needed

- an X-ray picture

1. An X-ray machine can take a picture of your bones. If you break one of your bones, the X-ray helps the doctor find the break. Help each child to look at the X-ray and identify the bones.
2. On the lab sheet, circle the picture of the boys and girls who are being careful not to break bones. Cross out the unsafe play.

Station Six

Materials Needed

- snack food (e.g., cheese and crackers, broccoli and dip, or a glass of milk)

Provide a calcium rich snack like milk or cheese and crackers at this station. Calcium helps to keep bones strong. Discuss the importance of eating foods rich in calcium.

Station Seven

Materials Needed

- Body Builder Card 6

Give the parent helper Body Builder Card 6. Follow the directions and have your children add the backbone to the body outline.

What Scientists Know

You have many bones in your body, and they all fit together to make your skeletal system. Besides holding you up, bones also protect some of your most important parts. You discovered that your spine is protected by your backbone. Did you know that your heart and lungs are protected by your rib bones? Your brain is also inside a bone called the skull. Bones are joined together at places called joints. Most joints move. Joints in your fingertips, at your elbows, and at the knees all help you to move. Bones are hard on the outside and sponge-like on the inside. This makes our bones very strong but very lightweight. The inside of your bones stores calcium and other good things that you eat. Not only do bones support us and protect us, they produce new red blood cells, as well.

Real Life Application

If you live by a university, call and ask the biology department if they have a real skeleton. Some high schools also have skeletons. Take a field trip to visit a skeleton.

Body Builders

If you set up all these centers in your room, you have completed Body Builders cards 3 and 6 to add bones and a backbone to your body.

Dem Bones *(cont.)*

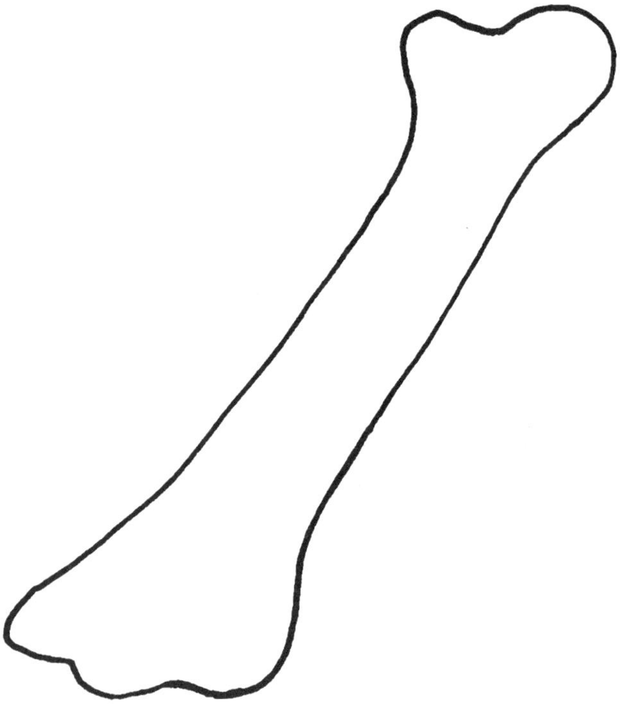

Draw the inside of this bone.

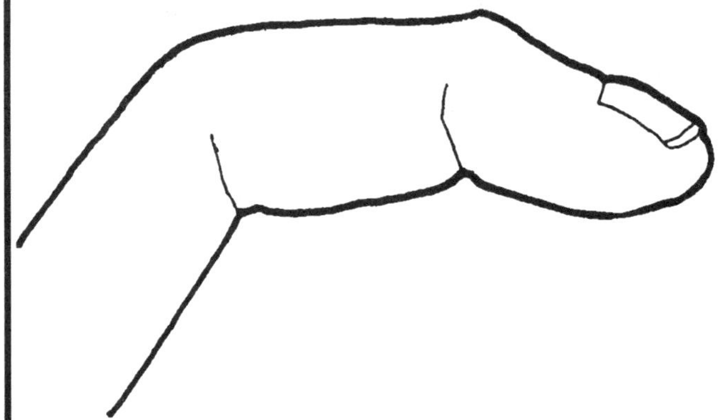

Circle the joints in this finger.

○ pictures of children playing safely

✗ pictures of children who need to be more careful

Dem Bones *(cont.)*

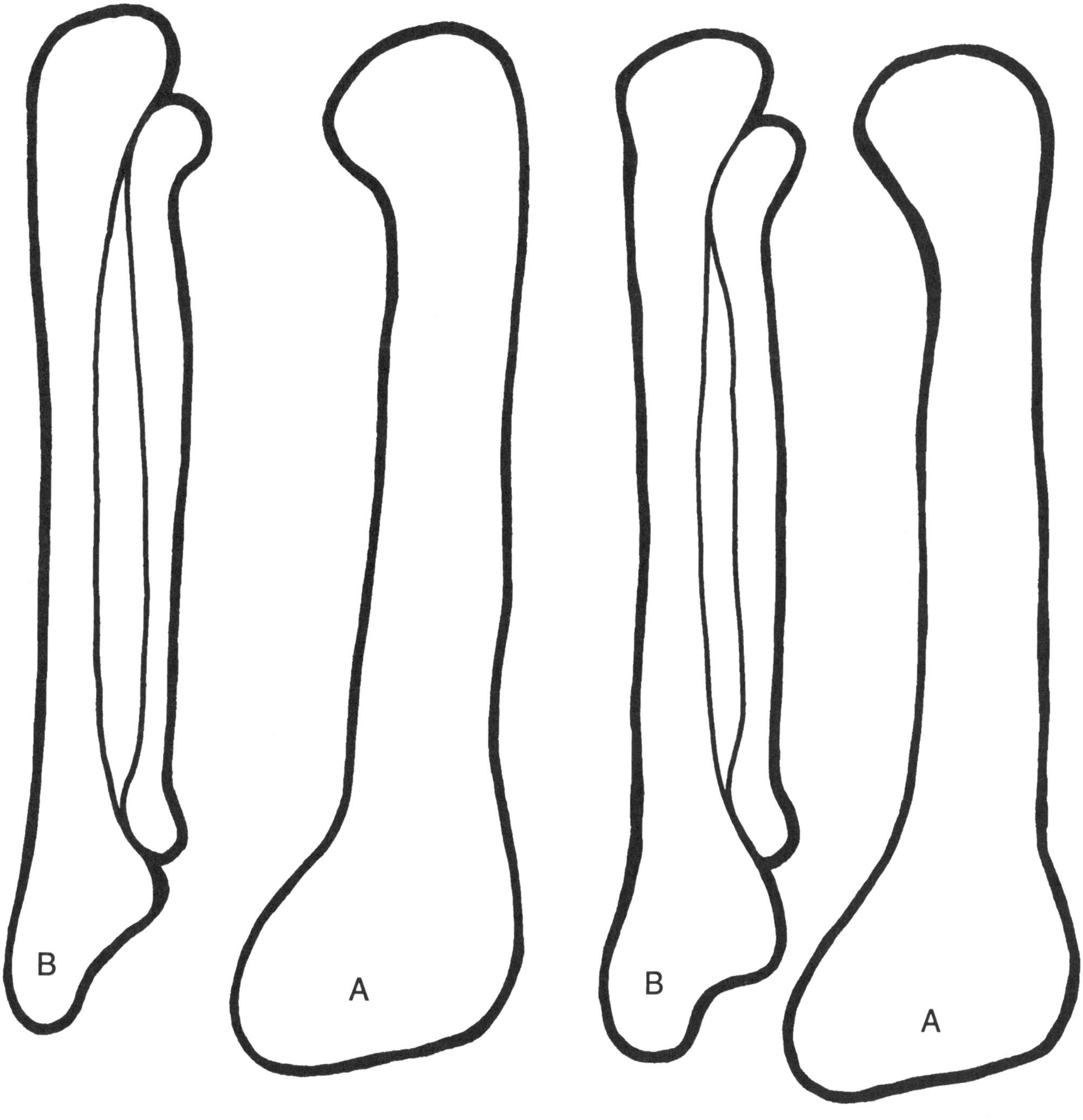

Dem Bones *(cont.)*

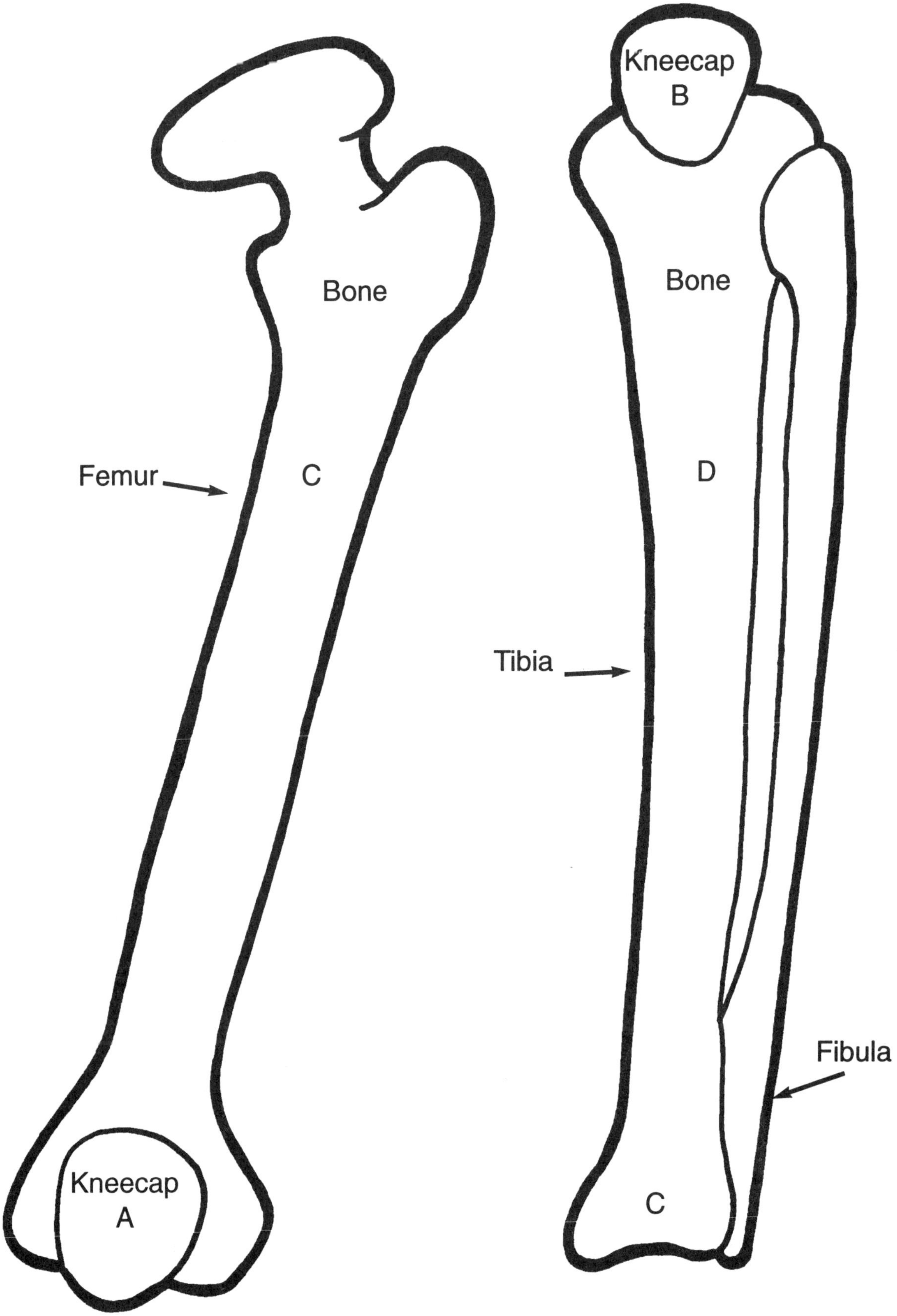

Movin' with My Muscles

Science Concept: Muscles help me to move.

Concept Literature: *You Can't Make a Move Without Your Muscles* by Paul Showers

Questions

What does muscle look like? What does muscle do?

Materials Needed for the Class

- a chicken leg with the thigh still attached (Keep the chicken chilled until you are ready to use it.)

Discovery Experience

1. Ask everyone to make a muscle.
2. Explain to your children that we have many muscles under our skin. Muscles help us move.
3. Show your children the chicken leg. Peel off the skin. (The white sheath, or tendon, just under the skin is a part of the continuous web of connective tissues that attaches bone to muscle.)
4. The meat we eat on a chicken leg is really the muscle of the chicken. Your muscle looks like this, too.
5. Look for places (tendons) where the muscle is attached to the bone.
6. By pulling on the tendon you can move the bone. Try this on the chicken leg. If the muscles move, bones move. If bones move, skin moves. If skin moves, you move.

What Scientists Know

Every time you smile or frown, jump, or read a book you use your muscles. There are 600 muscles in your body. Many of your muscles are attached to bone. When you want to wave to a friend, your brain tells the muscles in your arm to move. The top muscle in your arm bunches up (contracts). The muscle under your arm lengthens. This muscle action pulls your arm bone up. This in turn pulls your hand up, and you are able to wave.

Real Life Application

Lead your children in a set of exercises that use different muscle groups. For instance, run in place and talk about their large, strong leg muscles. Do arm circles until the arm muscles feel fatigued. Some muscles can get tired. When your children feel their hearts pounding, be sure to tell them that the heart is a muscle, too. Record each child's accomplishments on the next page, "I Used My Muscles Today," and send it home.

Body Builders

Add biceps muscles to the arms of your body outline and show your children how the muscles contract and lengthen. See card 7 for instructions.

Movin' with My Muscles *(cont.)*

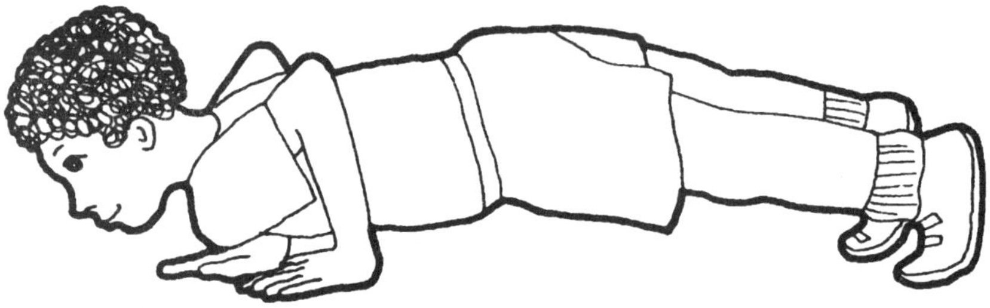

I Used My Muscles Today

I can do sit-ups. I did __________ sit-ups today.

I can do push-ups. I did __________ push-ups today.

I can do jumping jacks. I did __________ jumping jacks today.

Hurray for my muscles and me!

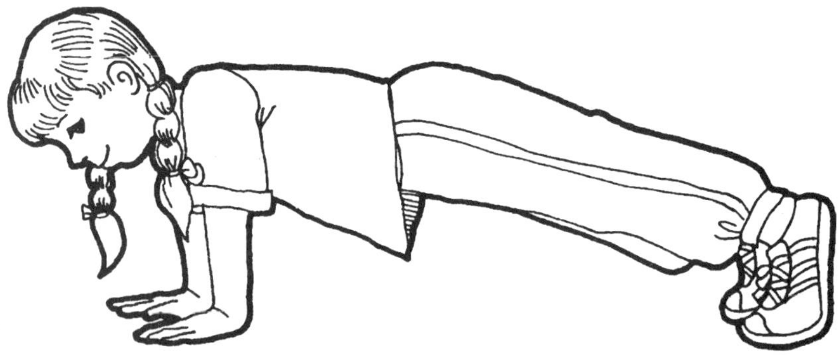

I Used My Muscles Today

I can do sit-ups. I did __________ sit-ups today.

I can do push-ups. I did __________ push-ups today.

I can do jumping jacks. I did __________ jumping jacks today.

Hurray for my muscles and me!

Sniff a Friend

Science Concept: My sense of smell tells me about the world.

Concept Literature: *Smelling Things* by Allan Fowler

Question

Can I use my nose to locate smells?

Materials Needed for Each Child

- 1 film canister (These can be obtained free of charge from a photo developer.)
- an assortment of strong-smelling items (e.g., coffee, cologne, cinnamon, vinegar, raw onion, rubbing alcohol, vanilla, etc.)
- cotton balls
- masking tape

Advanced Preparation

You will be dividing your class into four equal (or nearly equal) groups, so prepare enough film canisters of one scent for everyone in the group to have one canister with the same smell. For instance, if you have 20 children, you may choose to prepare five canisters with rubbing alcohol, five with cologne, five with vinegar, and five with coffee. Prepare the canisters by drenching a cotton ball in the substance or hiding the substance underneath a cotton ball. Snap the lids on each canister to preserve the odor. Mark like canisters with a secret symbol for easy checking later. Note: These canisters will retain their odors for several months if the lids are secure.

Discovery Experience

1. Before school, spray some air freshener or boil a little potpourri to give your classroom a pleasant scent.
2. As your children enter the room encourage them to talk about the odor they smell.
3. Tell your children the story about the amazing little ant. "Little ant was scurrying about in the tall, cool grass while looking for food. All of a sudden, another ant came charging towards little ant. Was this visitor a friendly ant or an enemy ant? The little ant reached out his antennae and smelled the new ant. He recognized the smell and knew this ant was from his nest. The two ants continued on, looking for food together." Ants communicate with smells. Ants from the same nest have the same smell. When little ant met the new ant he was not worried because he recognized his smell. If an ant is in danger, he can give off a warning odor to alert the other ants. Ants smell with their antennae. What do you use to smell?
4. Tell your children that, just like little ant, they will be using their noses to find a group of friends today. Mix up the canisters and distribute them to your children.
5. Allow each child to smell the contents of his/her canister.
6. They should then circulate around the room, smelling each other's canisters until they locate one or more children with the same smell. Continue the game until all are in their groups.
7. Can your group guess what smell is inside your canisters?

Sniff a Friend *(cont.)*

What Scientists Know

Inside your nose, just behind the bridge, is a smelling patch. This smelling patch is small, about the size of a postage stamp, but it can help you to smell about 10,000 different odors. The air you breath contains tiny smells (odor molecules). Every time you inhale through your nose, these smells rush over the smelling patch inside your nose. Special nerves on the smelling patch carry the smell signal to the brain, and your brain helps you to identify the odor.

Real Life Application

Take your class on a smell walk to "scents-a-tize" your children to a variety of smells. Visit the cafeteria just before lunch. Go outside after a rain storm. Visit the science lab. Make a quick stop by the trash dumpster. Your nose can tell you a lot about your world.

Complete page 3 of the *My Five Senses Tell Me About the World* Do-Along Book.

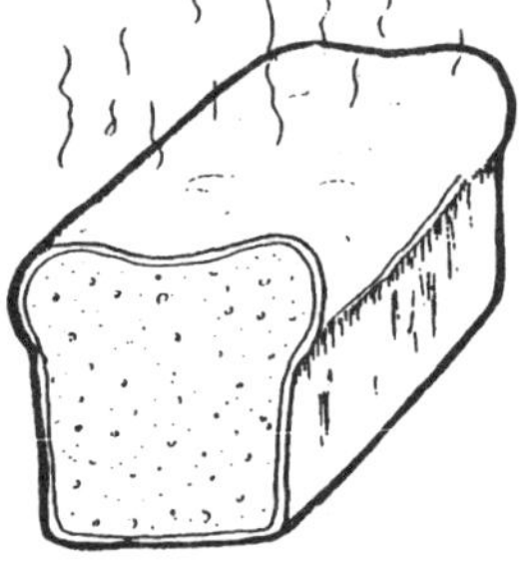

Huff and Puff

Science Concept: My lungs hold the air that my body needs.

Concept Literature: *Your Insides* by Joanna Cole

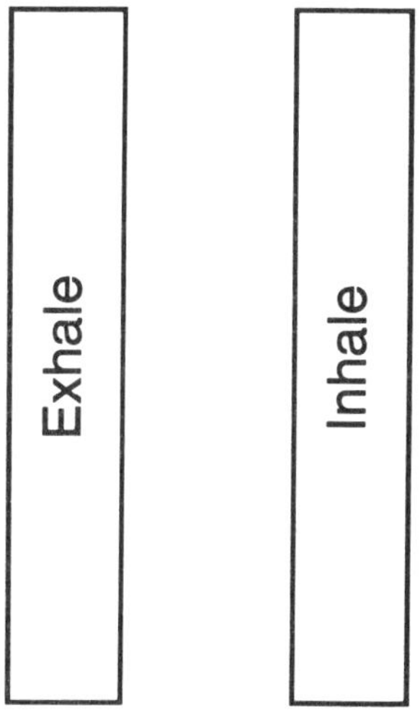

Question

What happens when I breathe?

Materials Needed for Each Child

- adding machine tape
- scissors
- red and blue markers or crayons
- masking tape

Discovery Experience

1. Besides smelling, what else do we do with our noses? Have each child take a deep breath through his/her nose and hold it for a moment. As they hold in their breath, ask your children to place their hands on the part of their body that is holding the air.
2. As your children exhale, have them keep their hands on their chests and describe what they feel. Have each child inhale and exhale several times. Encourage your children to feel the rhythm of their chests. Introduce the terms inhale and exhale.
3. Have a volunteer take a deep breath. Use the adding machine tape to measure the diameter of his/her chest during the inhale. Write inhale on this strip with a red marker as the child releases the breath. Hold up the adding machine tape up for all to see.
4. Ask the same volunteer to take in a deep breath through his/her nose and exhale as the class watches. Explain that in a moment you will measure his/her chest again as he/she exhales. Allow your children to predict how the length of the second exhale tape will differ from the length of the first inhale strip.
5. Show your volunteer how to exhale and force most of the air out of the lungs. Quickly use another piece of adding machine tape to measure his/her chest after he/she exhales. Write *exhale* on this strip with a blue marker.
6. Hold the two tapes side by side and have your children consider how they are different and why.
7. Measure everyone's inhale and exhale diameters. Children may wish to color their inhale strips red and the exhale strips blue so that each can be easily identified. Hang up the results along the wall as shown. Are everyone's measurements the same? Is the inhale strip always bigger than the exhale strip? Do you think some children can take and hold a bigger breath than others?

What Scientists Know

Each time you inhale, your chest expands to take in the air. Inside your chest are two balloon-like organs called lungs. Your lungs bring air into your body so that your body can use the oxygen. When you exhale, air leaves your lungs. As you grow, your lungs will be able to hold more and more air. Sometimes, the bigger children in the class can hold more air in their lungs than the smaller children. That is because their lungs are bigger.

Real Life Application

Give each child a generous piece of adding machine tape. Have your children measure their parent's inhale and exhale and compare it with their own.

Nose, Mouth, and Lungs

Science Concept: There are two passageways into the lungs.

Concept Literature: *Smelling* by Kathie Smith and Victoria Crenson

Question

How does air get into my lungs?

Discovery Experience

1. Have each child take a deep breath through his/her nose, hold it a moment, and then exhale through his/her nose. You can bring air into your lungs through your nose.
2. If your nose was completely plugged up, could you still get air into your lungs? Have each child pinch his/her nose shut. Can you still fill your lungs with air? How? Give your children time to discover step 3 on their own.
3. This time close your lips tightly. Breathe only through your nose. Lead your children to conclude that there must be two passages into the lungs.
4. Explain that it is possible to breathe in through the mouth or nose, but it is preferable to inhale through the nose. Air is dirty and often cold. The nose has tiny hairs inside that clean the incoming air. The nose also warms the air headed for the lungs.

What Scientists Know

Air can enter the windpipe (trachea) and then the lungs through either the nose or the mouth. This is because the mouth and nose share the same passageway (pharynx) for a few inches (cm) before going their own ways. A little flap (epiglottis) inside your throat keeps food from going into your windpipe, but choking can occur if we accidentally inhale food down the windpipe. That is why it is important not to laugh and talk when you are chewing.

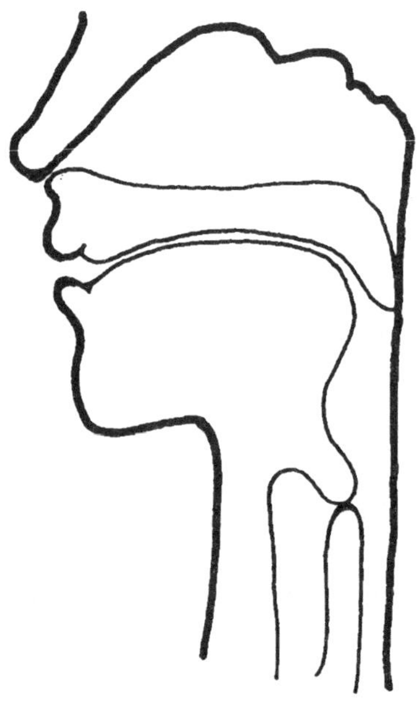

Real Life Application

The nose is like a vacuum cleaner. It sucks and it cleans. But is the air really dirty? Try this simple activity to show the children what is floating in the air. Give each child an index card. Smear a quarter sized dab of petroleum jelly on the card. Allow each child to tape his/her index card in a location of his/her choice. For instance, tape one inside the classroom. Tape one near the parking lot and another near a furnace or air conditioning vent. Label each card with the date and location as shown. After one week, gather the index cards. What is stuck in the petroleum jelly? Are some cards dirtier than others? Why? Explain once again that our bodies need clean air to function. The nose cleans and filters tiny particles that are floating in the air.

Special Delivery

Science Concept: Blood delivers oxygen from the lungs to all parts of the body via blood vessels.

Concept Literature: *A Drop of Blood* by Paul Showers

Question

How does the fresh air (oxygen) I breathe get to my fingers?

Materials Needed For Each Child

- their body

Discovery Experience

1. Have children share about "ouchies" they have had. Encourage your children to describe the blood that may have resulted from a cut or scrape.
2. Explain that blood is circulating inside their bodies right now and doing a very important job. It is delivering oxygen from the lungs to each part of the body. Show your children pictures of the circulatory system from a body book. Introduce the words, blood vessel or vein and share some of the information from, What Scientists Know below.
3. Ask them whether they can see a place under their skin where the blood is moving through vessels.
4. Try this activity to show your children the result of blood flow: Have each child stand and raise one hand high above his/her head and leave the other hand hanging at his/her side. Hold this position for one minute and then compare hands. The high hand will be pale, and the low hand will be darker. Why are their hands different colors? Allow children to share their ideas.

What Scientists Know

Blood travels through tunnels that are inside your body. These tunnels are called blood vessels. Some vessels, like the ones inside your eyes, are very tiny, while others, like the ones near your heart, are thick and muscular. If you could look inside your body you would see thousands of blood vessels running around inside every single inch (cm) of you. Blood moves through your vessels all the time, even when you are asleep. This is called circulation. Why is your blood so busy? Your blood delivers oxygen, which you need to survive, to all parts of your body. Your blood is the delivery service of your body.

Real Life Application

Have you ever sat with your leg folded underneath your body? After awhile your foot starts to tingle. It feels like someone is pricking your foot with a needle. A doubled up leg is like a garden hose with a twist in it. Just as water can only trickle through a twisted garden hose, blood runs slowly through a twisted leg. Blood not only delivers fresh oxygen to your foot, it also takes away the body's wastes. When the foot does not get a normal flow of oxygen-rich blood, waste can build up in the foot. The leftover wastes actually poison the nerve endings and cause a tingling in your foot. When normal blood-flow is restored, the tingling disappears.

Body Builders

Add a heart and blood vessels to your body. See card 9 for instructions. Show your children how blood circulates from the heart to all points of the body.

My Heart Is a Pump

Science Concept: The heart is a muscle that pumps blood around and around inside the body.

Concept Literature: *The Stethoscope Book and Kit* by Linda Allison and Tom Ferguson

Question

How does my heart function?

Materials Needed for the Class

- 3 large, heart-shaped paper charts titled: *What I Know About My Heart, Questions I Have About My Heart,* and *I Learned Something New About My Heart*

Note: Additional materials are listed below.

Discovery Experience

1. Introduce the first paper heart—*What I Know About My Heart.* Have your class brainstorm what they already know about the human heart. Record their ideas, including both accurate ideas and misconceptions, on the paper heart.

2. Introduce the second paper heart—*Questions I Have About My Heart.* Ask your children if they have any questions about the human heart. Record the questions and then keep this heart handy so that any additional questions may be added as they arise. Over the next several days, try some or all of the following activities designed to give your children a better understanding of their hearts. If materials are needed, they are listed with each activity.

Activity 1—How Big Is My Heart?

Have each child make a fist. Your heart is about as big as your fist.

Activity 2—Where Is My Heart?

Have each child make a fist and place it in the location that he/she believes his/her heart to be. Some children may be surprised to discover that their hearts are actually in the middle of their chests, between the lungs and behind the ribs. Pause to let each child feel the beating of his own heart.

Activity 3—What Does a Heart Look Like?

Materials Needed

- a picture of a human heart

Compare one of the valentine heart charts with a picture of a human heart. How do they differ? (Note: Some butchers will donate animal hearts to teachers for classroom use.)

My Heart Is a Pump *(cont.)*

Discovery Experience *(cont.)*

Activity 4—How Does My Heart Pump Blood?

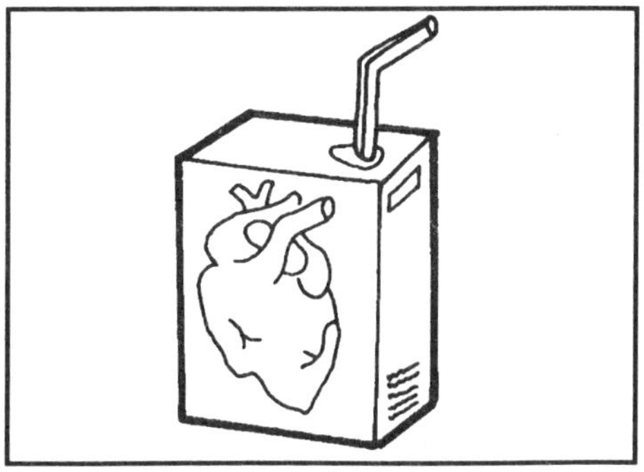

Materials Needed for Each Child

- a juice box containing a red drink
- a copy of the juice-box heart from page 75
- a glass
- crayons
- transparent tape

Have each child color the heart from page 75. Wrap the heart around a juice box and tape it in place. Explain that the heart is a pump. It pumps by squeezing. Allow each child to puncture his/her juice box with the straw. Place a cup underneath the straw and then have each child squeeze the box. Each time the box is squeezed red juice will come squirting out of the straw, illustrating how the heart works. Compare the squeeze to the squeeze of the heart, the straw to an artery, and the red juice to the blood that is pumped from the heart with each squeeze. The heart is a muscle that pumps blood.

Activity 5—Tell Me More About How My Heart Pumps

Materials Needed

- a long red balloon

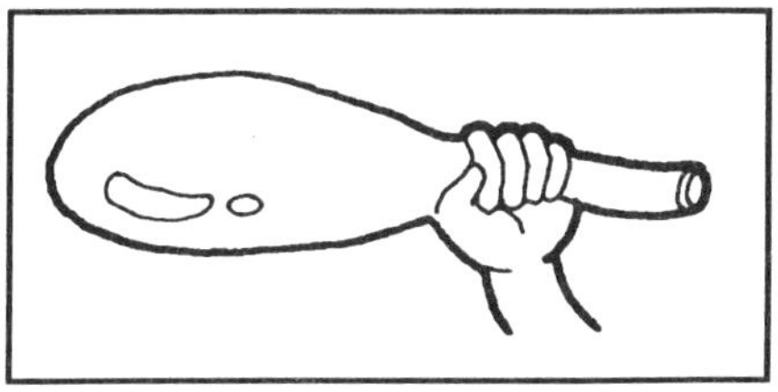

Partially inflate a red balloon and tie it shut. Show your children that by squeezing on one end of the balloon they can force all the air into the other side of the balloon. Explain that the heart moves blood through the body with this same type of squeezing power.

Activity 6—How Often Does My Heart Muscle Pump?

An adult's heart beats about 70 times every minute. Try opening and closing your hand with a squeeze to make a fist 70 times in one minute. How does your hand feel? Your heart muscles squeeze like this every minute of every day, but they do not get tired.

Activity 7—Can I Make My Heart Pump Faster?

Materials Needed for Each Child

- Body Builder Chart 13

Have each child feel his/her resting heart beat with his/her hand. Next, have your children run in place and then place their hands over their hearts again. Can you feel your heart pounding? Why does the heart work harder when you run? Read Body Builder Card 13.

Activity 8—Why Do I Feel a Beat in My Heart and in My Neck?

Help each child find a pulse on the side of his/her neck. Allow your children to get a feel for the resting rhythm of the pulse. Lead your class in a rousing chorus of "Head and Shoulders, Knees and Toes." Quickly have each child find the pulse again in the same spot. Does the beat feel stronger and quicker? Help your children to make a connection between exercise and oxygen and heartbeat and the circulatory system.

My Heart Is a Pump *(cont.)*

Discovery Experience *(cont.)*

Activity 9—How Does My Heart Sound?

Materials Needed for Each Child

- paper towel tube

Give each child a paper towel tube and allow him/her to listen to a friend's heart. The lub-dub sound you hear is the sound of "doors" (valves) inside the heart closing as the blood moves in and out of the heart.

Activity 10—Where Is All the Blood Going?

Materials Needed for the Class

- a tunnel

If you have a tunnel on the playground or inside your classroom, you can play this game. Pretend that you are the heart, your students are the blood, and the tunnel is an artery. Stand at one end of the tunnel with a child. Give the child a big hug to represent a pump of the heart and then send him/her through the artery to a place inside the body. For instance, you may give the child a hug and say, "I am the heart, and I am sending you, the blood, through the artery and then to the big toe. Go blood! Go!" The child then crawls through the artery and pretends to head towards the toe. Children will circulate around again and again as you send them all over the body with a squeeze.

3. After trying several of the activities listed above, post the final paper heart—*I Learned Something New About My Heart.* Take dictation on the last heart-chart as your children list new things they have learned. **Note:** With very young children you may wish to add to this list immediately after completing each activity.
4. Look back to the second heart-chart—*Questions I Have About My Heart.* Have all the questions been answered through the hands-on activities? If not, take this opportunity to model a love of books by reading and researching to locate information.
5. Look back to the first heart chart—*What I Know About My Heart.* Were there misconceptions that have been corrected? Gently lead your children to draw new conclusions.

What Scientists Know

Your blood moves (circulates) inside your body all the time. You do not need to think about it or remember to tell the blood to stop by your big toe with some oxygen. That is the job of the heart. The heart is like a big pump. Each time the heart fills up with blood it knows what to do. Squeeze! Just like the toothpaste in the tube, a squeeze sends oxygen-rich blood rushing through the veins. In fact, blood can travel from your heart to your big toe and back again in about the time it would take you to slowly count to 62. Your heart pumps blood day and night so that your body has all the oxygen that it needs. When you exercise or move quickly, your body needs more fuel (oxygen). Your heart responds to exercise by pumping faster to supply the entire body with more oxygen-rich blood.

Real Life Application

Invite a nurse to be a special guest. Ask the nurse to explain the basics of keeping a healthy heart. Have the nurse bring a stethoscope and allow your children to listen to their hearts.

My Heart Is a Pump *(cont.)*

Juice Box Hearts

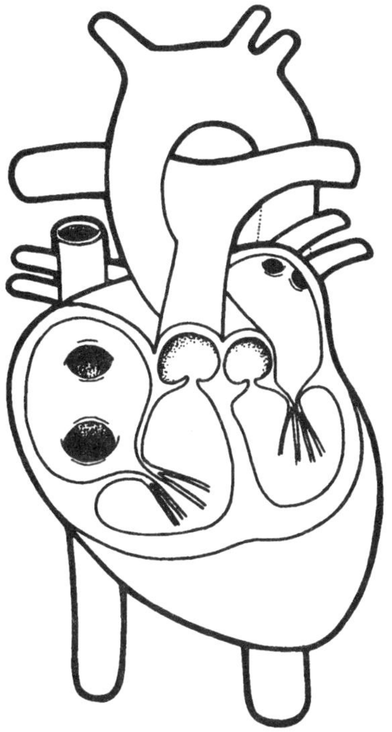

FOLD

The HEART is a PUMP

FOLD

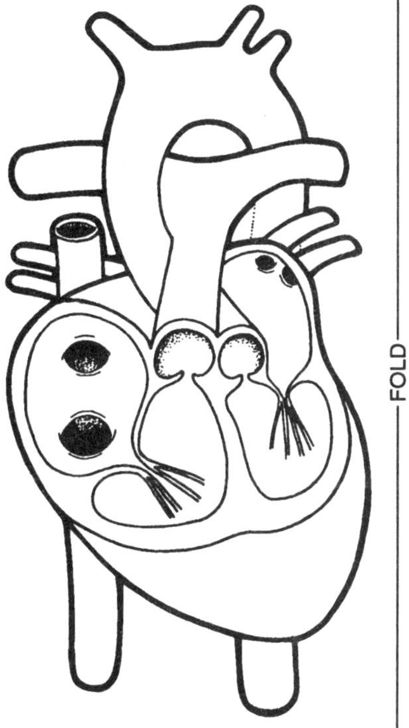

FOLD

The HEART is a PUMP

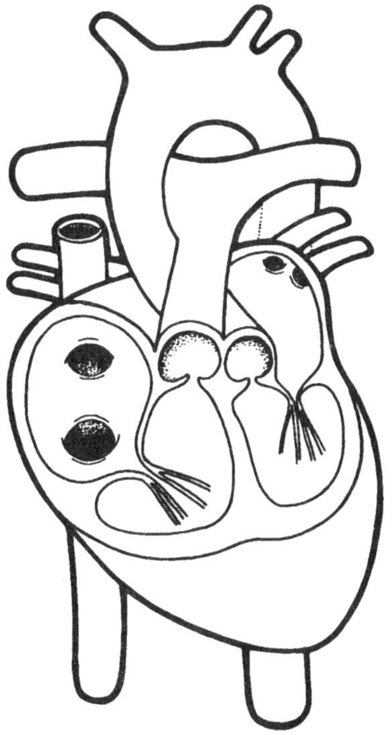

FOLD

The HEART is a PUMP

FOLD

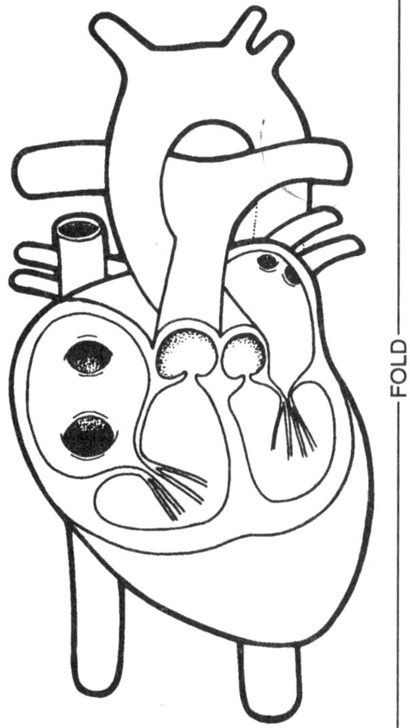

FOLD

The HEART is a PUMP

Teachers: Cut on the dotted lines and give each child a heart. Color the heart. Wrap it around a juice box filled with red juice and tape it to the juice box. Use this juice box heart with Activity 4. "How does my heart pump blood?"

Tasting Is Believing

Science Concept: My sense of taste tells me about the world.

Concept Literature: *Tasting Things* by Allan Fowler and *A Tasting Party* by Jane Belk Moncure

Question

How can I identify two substances that look similar?

Materials Needed for Each Child

- nut or condiment cups
- granulated sugar
- salt
- 1 cup (250 mL) of water
- 1 small cracker
- clean hands

Discovery Experience

1. Give each child a small cracker. Ask your children to take a bite of the cracker.
2. What did you just eat? How did you know it was a cracker?
3. Give each child two nut cups, one containing a teaspoon of sugar and the other a teaspoon of salt. Do not reveal the identity of the white substances.
4. Ask volunteers to look carefully at the white substances and guess what they are. Are the white substances the same or are they different? Can you tell just by looking what each substance is? How could you find out for sure what is in the nut cups?
5. Have each child lick a finger in order to taste and identify each substance. A sip of water in between will help to clear the palate.
6. Help your children to discuss how their sense of sight was not reliable in identifying the white substances. You had to use your tongue and your sense of taste in order to correctly identify the white substances.

What Scientists Know

The tongue sends messages to the brain every time you take a bite. Your tongue is covered with tiny tasters called taste buds. There are four different kinds of taste buds: sweet taste buds and sour, bitter and salty taste buds. Your tongue is your most important taster, but your eyes and nose help to tell your brain what is coming, too. That is why a stuffy nose can make food taste blah.

Real Life Application

I like mine sweet. I like mine salty. Not everyone has the same taste preferences. Children need many opportunities to explore different foods and different tastes. Pop up a batch of sugar popcorn and a salty batch of regular popcorn. Enjoy each type of popcorn and talk about sweet and salty. Then, make a class graph, like the one shown, to illustrate taste preferences. Duplicate and cut out the markers on the next page and have each child use one marker to vote for his/her favorite popcorn flavor.

Complete page 4 of the *My Five Senses Tell Me About the World* Do-Along Book.

Mouth Watering Science

Science Concept: Digestion begins in the mouth.

Concept Literature: *What Happens to a Hamburger?* by Paul Showers

Question

What happens to food inside of my mouth?

Materials Needed for Each Child

- 1 Sweettart
- 1 raw carrot stick
- 1 paper plate

Discovery Experience

1. Give each child a Sweettart to hold in his/her hand. Some children who are familiar with Sweettarts may find that holding the candy and looking at it causes them to salivate. What is happening inside your mouth? Allow everyone to lick the Sweettart and continue the discussion. Help your children to understand that the mouth salivates in anticipation of food.
2. Give each child a carrot stick and a paper plate.
3. Ask your children to bite the carrot, chew it five times, and then spit it out onto the paper plate. Observe the chewed carrot. What is mixed in with the carrot? How has the carrot changed?
4. Take another bite, chew it 20 times, and spit it out next to the first sample.
5. Compare the two mixtures. How are they different? What is happening inside your mouth? How does saliva help us to swallow? What have your teeth done to the carrot?

What Scientists Know

As soon as you put food into your mouth, your body begins to break it down so that it can be used. Your teeth grind the food into tiny bits that are easy to swallow. Spit or saliva flows into your mouth and mixes with the food. Saliva helps to break down food and make it mushy so it is easy to swallow. Have you ever tried to eat a cracker without chewing it correctly? It may feel hard and dry when you try to swallow it. You may even choke if you do not chew your food thoroughly. Chewing and saliva work together in order to prepare the food to go inside our bodies.

Real Life Application

During snack time or lunch remind your children that it is important to take small bites and chew food slowly and thoroughly.

Going Down

Science Concept: After I swallow food, it is squeezed through the "going-in food tube" that leads to my stomach.

Concept Literature: *Food and Digestion* by Steve Parker

Question

How does food move from my mouth to my stomach?

Materials Needed for Each Child

- 1 piece of banana
- 1 glass of milk or juice
- clean hands
- 1 "tummy-phone" for every other child (directions below)
- 1-13" (32.5 cm) long, thin balloon

Discovery Experience

1. Lay out a piece of banana and a drink for each child.
2. Review the job of teeth and saliva from the previous activity, *Tasting Is Believing.*
3. Ask each child to take a bite of a banana and chew it eight times. As they chew, demonstrate how each child can place a hand on his/her throat in order to feel the food moving into the "going- in food tube."
4. Have each child swallow the banana and then feel it move down the esophagus. Have them describe what they feel. Where is the banana going? Have everyone finish the snack and continue to feel the swallowing process.
5. Ask each child to point to their tummies or the place where the banana is now. Most children will locate their stomachs at their navels. Show your children that the stomach is really much higher; it is located under the bottom ribs. Guide your children in a discussion about how they believe the piece of banana got into their stomachs.
6. Show your children a long, deflated balloon and place it on your throat over your esophagus. Explain that everyone has a food tube inside that is 10" (25 cm) long and thin like this balloon. This tube that connects the mouth and the stomach has an important job. It squeezes food into the stomach.

What Scientists Know

The banana did not just fall from your mouth into your stomach. It was squeezed down by muscles inside your throat (esophagus). The banana took only a few seconds to go down the "going-in food tube" and into the stomach. Once in your stomach, the stomach squeezes and shakes and churns and mixes in special stomach juice until the banana is as mushy as soup.

Real Life Application

Listen to a friend's stomach. By putting a "tummy-phone" on a classmate's stomach, you will be able to hear his/her stomach churning and mixing in stomach juice in order to turn the banana you ate into banana soup.

To make a "tummy-phone," punch out the bottom of a Styrofoam cup. Place this end on the stomach. Listen on the other end.

Body Builders

Add the "going-in food tube" and a stomach to your body outline.

See Body Builders card 10 for directions. Your children will get a feel for food moving into the stomach.

Moving Out

Science Concept: As food is digested it continues to move from the stomach into a long tube called the intestines.

Concept Literature: *Everyone Poops* by Taro Gaomi and *Your Insides* by Joanna Cole

Materials Needed for Each Group

- one pair of pantyhose cut as shown
- one banana
- one plastic bag
- transparent tape
- green food coloring

Discovery Experience

1. Ask your children what they think happens to food after it leaves their stomachs. Encourage children to share their ideas.

2. Show your children the general location of the intestines by coiling a pantyhose leg and holding it under the stomach on the sample body outline or on a child's body outline. Explain that after the food leaves the stomach, it passes through a long tube called the intestines. You may wish to show your children pictures of the digestive system.

3. To help your children understand that the intestines continue the mashing and souping digestive process that began in the mouth, try this activity, but be prepared for "Ugh!" and "Gross!"

4. Ask your children to recall how they ate a banana in the previous activity. Produce one banana, peel it, and be sure each child gets a good look at the firm, solid banana. Place the banana in a sandwich bag. Add a drop of green food coloring to represent digestive juices. Roll the bag around the banana and seal it with two pieces of tape. Be sure to squeeze all of the air out of the bag and wrap the tape around the bag to strengthen it. (Note: While it is good to prepare at least one bag in full view of your children, have the other bags prepared beforehand to save time. Folded sandwich bags are preferable as sealable plastic bags tend to burst inside the pantyhose.) Line your children up as shown and lay one pantyhose leg in their hands. Have them stretch the leg so everyone gets a bit of it. Place the bagged banana inside one end of the pantyhose. Show your children how to use their hands to push the banana along inside the bag. Each child should give the banana a squeeze to keep it moving. Continue to move the banana along until it appears at the opposite end of the pantyhose. Take the banana out of the bag and unwrap it to show your children your version of digested banana soup.

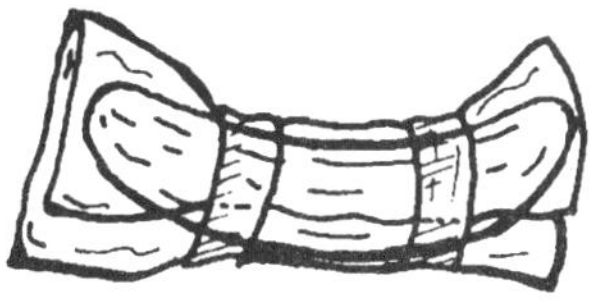

Moving Out *(cont.)*

What Scientists Know

Food leaves your stomach and goes into a long tube called the intestines. Inside the intestines food is mixed with body juices (digestive juices) and mushed up into very small bits that are even smaller than the bits of banana that came out of the pantyhose. To keep the food moving, muscles inside your intestine contract and relax (peristalsis) just as you kept the banana moving by squeezing and pushing it along. Your body needs food to give it energy and keep it healthy. While food is inside the intestine, your body picks out what it needs and leaves what it cannot use. Then the blood delivers the tiny bits of food to all parts of your body. After your body uses all it can of a food like a banana, it gets rid of the parts it does not need. This happens every time you go to the bathroom.

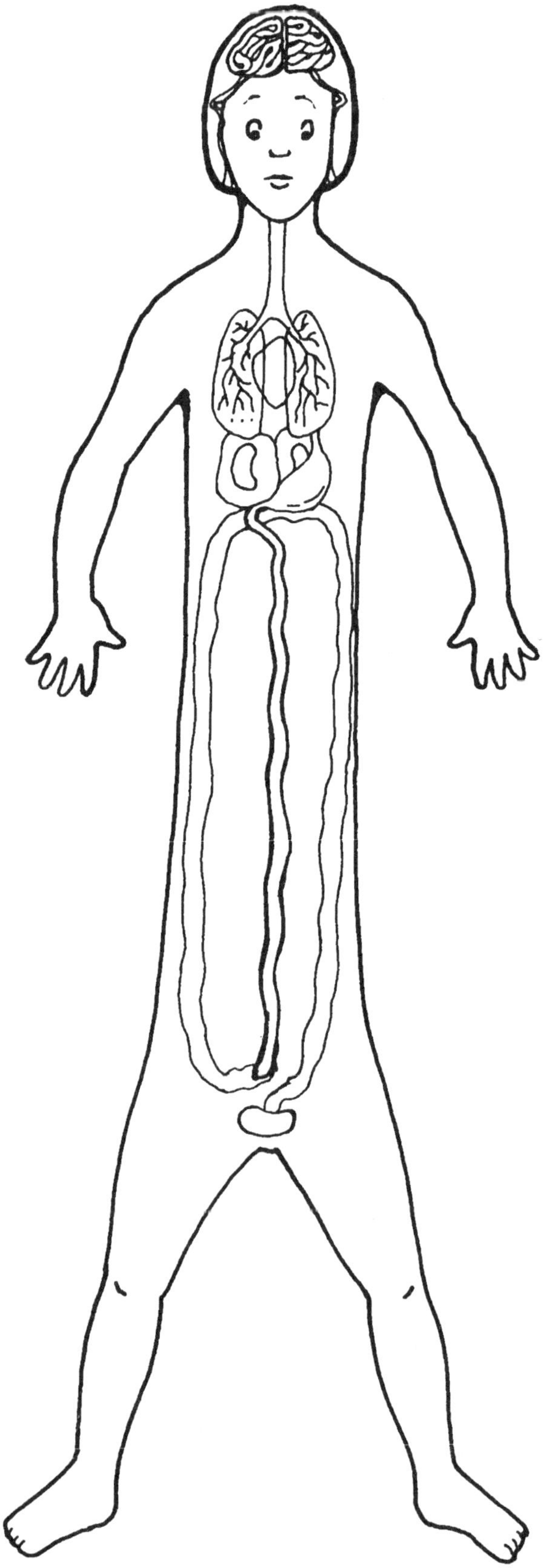

Real Life Application

Allow children to ask questions and share ideas about "poop." Children arc fascinated by this process. Begin the discussion with questions like: Does your pet poop? If there is a baby in your house, how is his/her poop different from yours? Why would you not want to keep all the food you have eaten inside your body? Why is it important to wash your hands after you use the bathroom? After going through some or all of these questions, read *Everyone Poops* to your children.

The digestive tract is really much longer than a pair of pantyhose. Help your children measure out approximately 20' (6.5 m) of adding machine tape. This is the actual length of the digestive system. It is amazing that it can all fit inside your body.

Body Builders

Add the intestines to your body. See Body Builder Card 11 for instructions.

Making Predictions Without My Eyes

Science Concept: My sense of sight tells me about the world.

Concept Literature: *Sight* by Maria Rius, J.M. Parramon, and J.J. Puig

Question

How do my eyes help me to make more complete observations?

Materials Needed for Each Group

- a yellow-skinned apple
- a brown grocery bag
- crayons, several different colors (Be sure to include red and yellow.)
- 1 additional apple, any color, for the entire class
- activity sheet, page 83

Discovery Experience

Before your children arrive, put one apple in each bag. Staple the bag shut several times. Cut an additional apple and rub the flesh on the outside of the bag to give the bag an apple smell.

1. As your children arrive, put the prepared bag in the middle of the table. Have each child place his/her hands in his/her lap and tell him/her not to touch the bag. Tell your children you want them to act like scientists today and guess what is in the bag on their table. Ask them "What could you do to the bag in order to make the best guess possible about what is in the bag?" Encourage anyone who says he/she could use one of the senses to gather data about the contents of the bag.
2. Pick up one of the bags and take it around the room so your children can smell its apple scent. Ask them, "After using your nose and your sense of smell, what do you think is in the bag?" Acknowledge all guesses but do not comment.
3. Next, allow one child in the group to pick up the bag at their table and shake it. Ask them, "After using your ears and your sense of hearing, what do you think is in the bag?" Again, acknowledge all guesses but do not comment.
4. Have each child pass the bag around the table and allow them to feel the apple through the bag. Continue the questioning pattern, emphasizing the sense of touch.
5. Ask your children to list the senses they have used so far to gather ideas about what is in the bag. "Which senses have you not used?"
6. By now your children will be confident that there is an apple in the bag. Pass out the activity sheet, page 83, and have them draw and color a picture on the space inside the bag at the top of the activity sheet that looks exactly like the item they believe to be in the bag. Most children will draw a picture of a red apple.
7. Finally, select one child in each group to open the bag. Your children will be pleased that they guessed correctly. Hold up several drawings next to the yellow apple. Ask them whether they were able to draw a picture of an apple that looks exactly like this one. Help your children to realize that without using their sense of sight they could not have known that it was a yellow

Making Predictions Without My Eyes *(cont.)*

Discovery Experience *(cont.)*

8. Have each child draw a yellow apple in the space provided on the activity sheet.
9. Slice the apple and have your children use the last of their five senses to observe the apple. When the lesson is over, allow your children to eat the apples.

What Scientists Know

To make complete observations, scientists use all of their five senses. The brain takes the information gathered by the five senses and comes up with guesses and new ideas.

Real Life Application

Highlight the eyes and sense of sight by playing a game of "I Spy" with your children. Instruct children to sit in a circle and play the "I Spy" game. Say, "My eyes spy something..." and then describe something your children can easily see in your room. For instance, if you were describing the American flag you might say, "My eyes spy something red, white, and blue and made of cloth." When your item is found, allow the children to spy items and describe them in clue form to your room of guessers. Use a piece of chart paper or the chalkboard to write down or draw pictures of the items as they are found.

Complete page 1 of the *My Five Senses Tell Me About the World* Do-Along Book.

Making Predictions Without My Eyes *(cont.)*

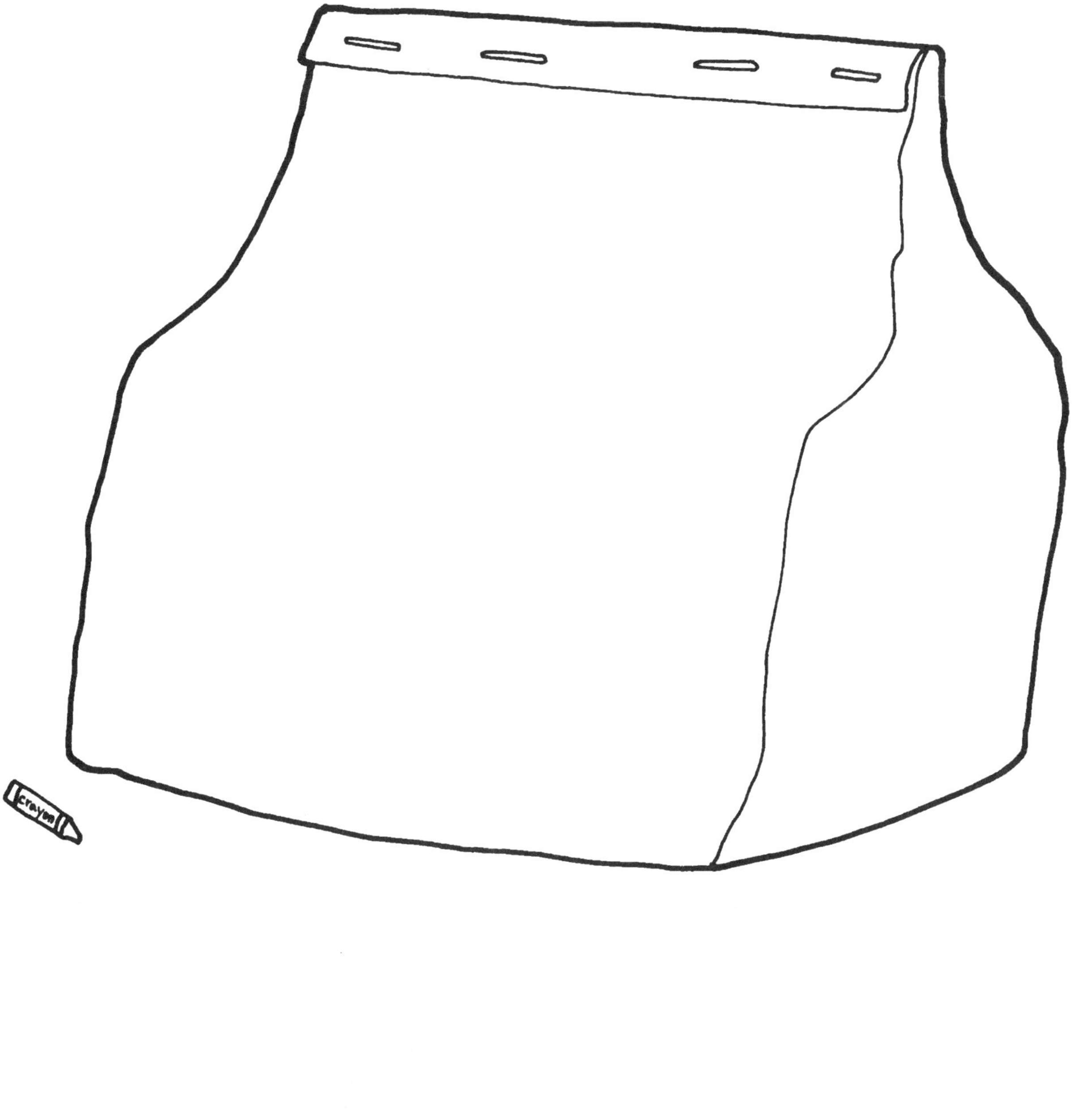

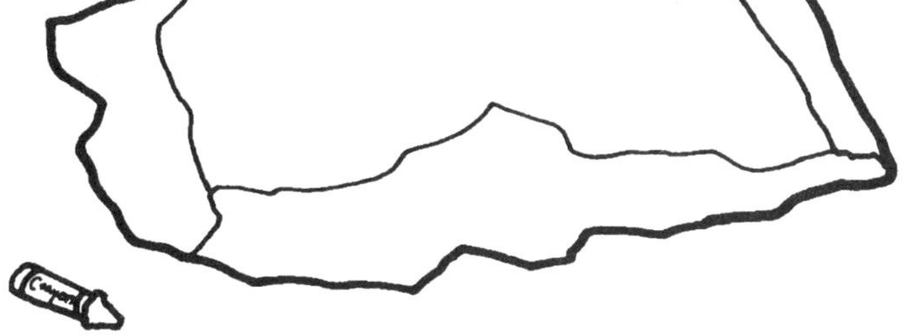

Behind Your Eye

Science Concept: The eye is more than what you see by just looking in a mirror.

Concept Literature: *What's Inside My Body* by Angela Royston

Materials Needed for Each Child

- hand mirror
- plastic sandwich bag
- large bottle of styling gel (One 16 oz. [500 mL] bottle is enough for about 12 eyes.)
- permanent marker
- pipe cleaner cut into three pieces
- large sequin
- glue
- wax paper square (size of a postage stamp)
- twist tie

Discovery Experience

1. Have children work in small groups for this activity.
2. Have children use the hand mirrors to examine their own eyes. Guide them in a discussion about the appearance of their eyes. Help children identify and name the outward features of their eyes.
3. Show children a simple picture of the inside of an eye. The parts in this simplified diagram are age-appropriate and can help you decide what needs to be emphasized.
4. Have children close their eyes and GENTLY press on their eyelids. They should be able to feel the shape of their eyes and how they sit in their skulls.
5. Have children make a model of the eye by following the directions on the following page.

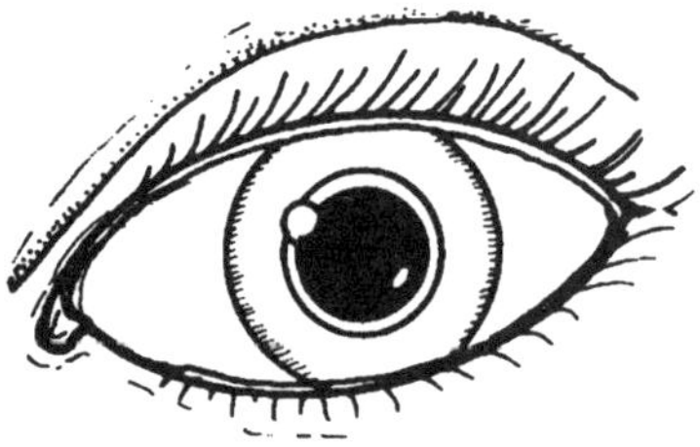

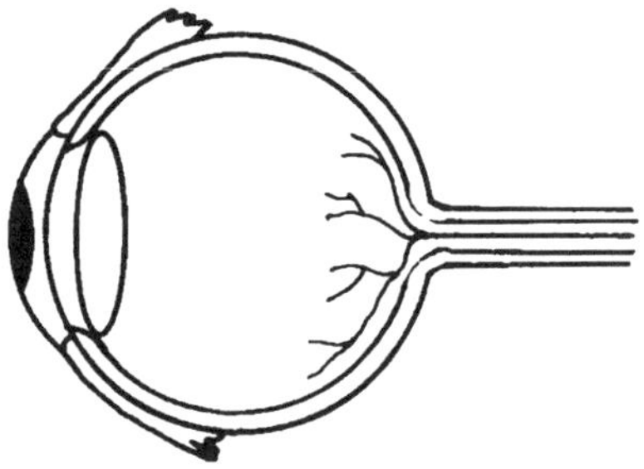

What Scientists Know

Look at a pencil. Light bounces off of the pencil and into your eye. The light passes through your eye (pupil and lens), and then the lens sends a sharp picture of the image (pencil) to your retina at the back of your eye. Nerves in your retina carry a picture message to your brain, and your brain recognizes this as a pencil. This happens so quickly you are unaware of it even occurring.

Real Life Applications

Review eye safety rules with your children. 1) Be careful to keep sharp objects away from your eyes. 2) When looking at books, be sure there is enough light to read. 3) If dirt gets in your eye, do not rub it. Let your tears wash it out. If your tears do not flush it out, ask an adult for help. 4) If you notice things are looking fuzzy, tell your parents. You may need to have your eyes tested to see whether you need glasses. 5) Visit an eye doctor once a year to keep your eyes healthy and happy.

Behind Your Eye *(cont.)*

1 Draw and color an iris and pupil on the outside of a plastic bag.

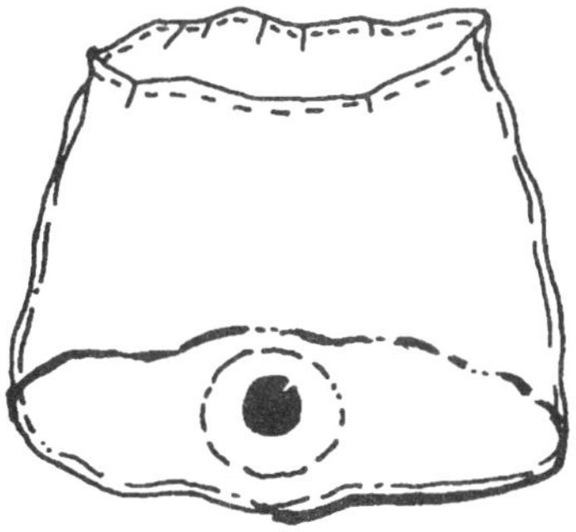

Use a permanent marker.

2 Glue a sequin inside the plastic bag, right behind the pupil.

This is the lens so the eye can focus.

3

Let the glue dry for several minutes.

4 Fill the plastic bag 2-3 teaspoons (10-15 mL) of hair gel.

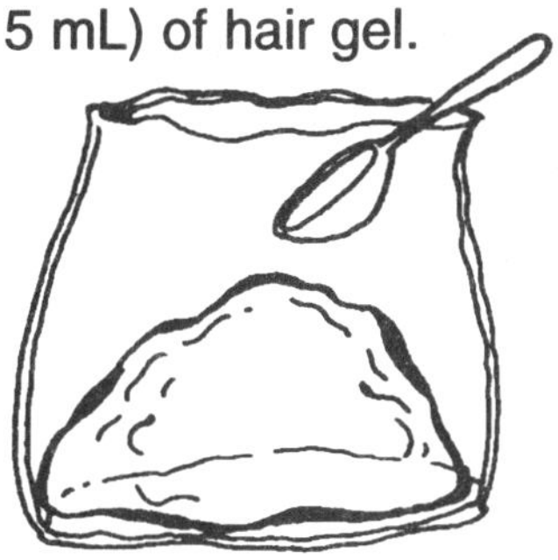

Put it right on top of the lens.

5 Keep the gel in the center of the plastic bag and gather up the top and sides.

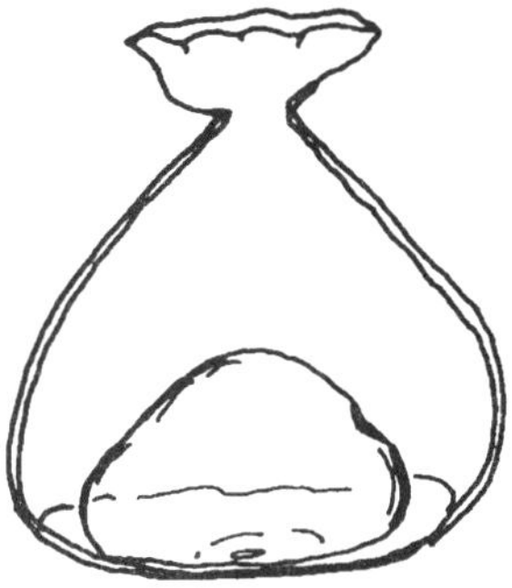

6 Stick a piece of wax paper on a pipe cleaner.

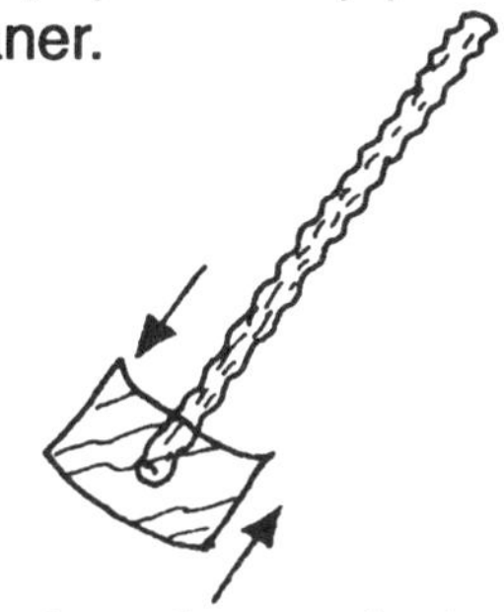

The pipe cleaner is the nerve and the wax paper is the retina.

7 Insert the pipe cleaner into the gel.

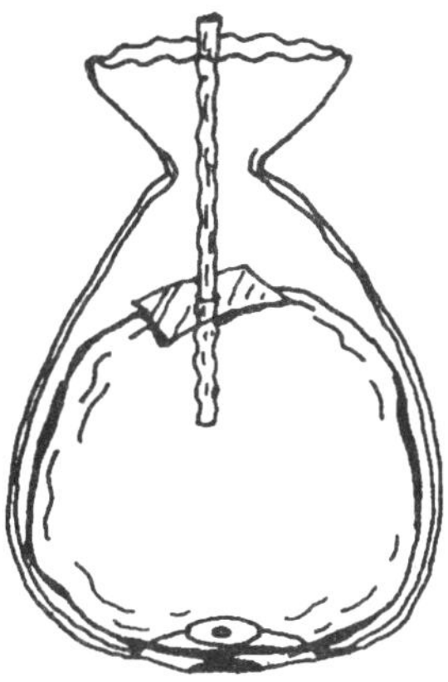

8 Use a twisty tie to close the plastic bag and to hold the pipe cleaner in place.

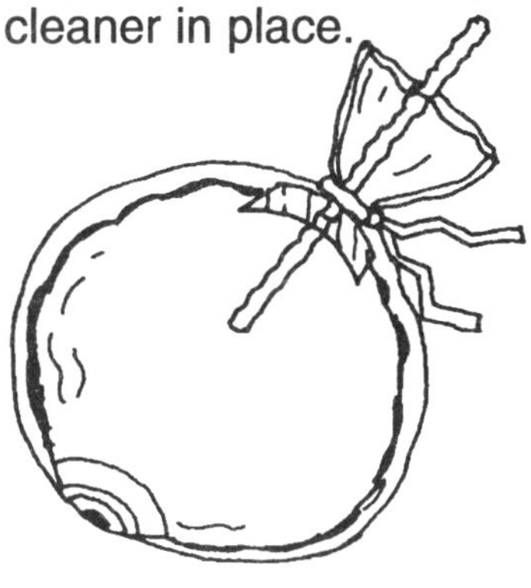

9 Cool! You have made an eyeball.

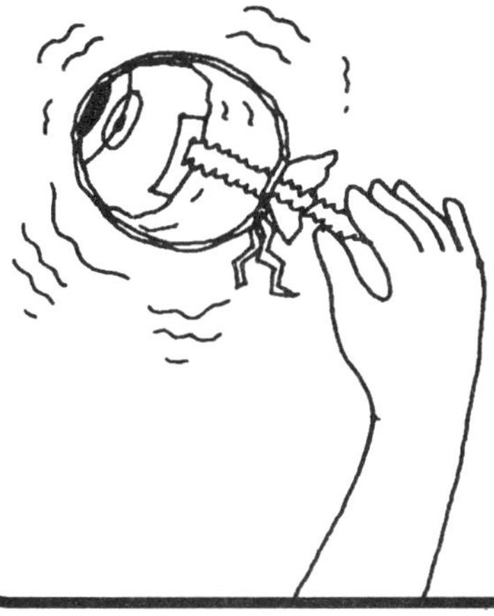

Put on Your Listening Ears

Science Concept: My sense of hearing tells me about the world around me.

Concept Literature: *The Other Way to Listen* by Byrd Baylor

Question

Can I use my ears to decide what is in a box?

Materials Needed for Each Group

- 3 to 5 shoe boxes (depending on the size of your class)
- 1 tennis ball
- 1 bell
- 1 pencil
- 1 cotton ball
- 1 handful of paper clips

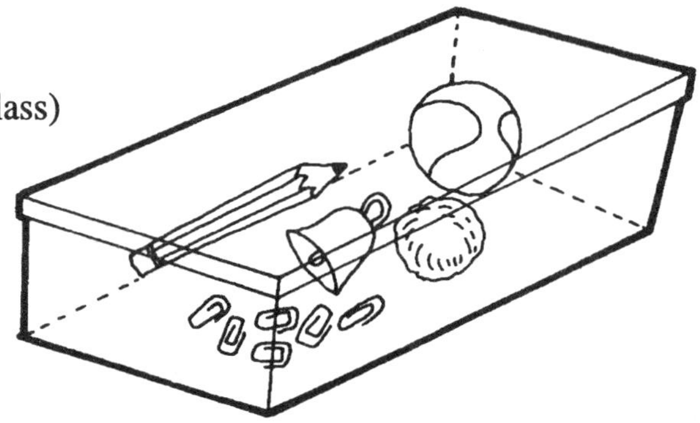

Discovery Experience

You will want to divide your class into groups of four for this activity. If you have a very large class, it is better to make groups of five than to add another group. Prepare the shoe boxes. First, put one of the items listed above inside each shoe box. You may wish to wrap the shoe boxes so they all look the same. Wrap the lid and box separately so that you can remove the lid. Securely seal the lid with tape.

1. Have your children sit at tables or on the floor in groups of about four. Place a sealed shoe box in the center of each group. Tell your children that you want them to try to guess what is inside the boxes. Have your children pick up their box and pass it around the circle. Encourage each child to shake the box and listen closely to the contents. Emphasize that they are using their ears to gather clues.
2. Encourage the groups to come to a consensus about the contents of their boxes. Have each group announce their guess one at a time and then open their box. Were they correct?
3. After all the boxes have been opened and their contents revealed, seal them again. Mix up the boxes and give each group a box they did not have in the first round. Repeat steps one and two. Was it easier to guess the contents of the box after seeing what was inside them all?

What Scientists Know

Hearing is important to you. You can hear me right now. You hear your mother call you for dinner. You hear cars going by on the street. You hear the alarm clock in the morning. If you are a good listener, you can hear many things that help you to learn about your world.

Real Life Application

Teach your children this chant and use it often to help children safely cross the street:

Stop, look and listen
Before you cross the street.
First, use your eyes and ears,
Then, use your feet!

Complete page 2 of the *My Five Senses Tell Me About the World* Do-Along Book.

Introduction to Body Builders

Would it not be helpful if, while studying the body, your children could look at life-size models of their own bodies? Better yet, what if they themselves could put one together? With Body Builders, they can. Children learn by doing, and that is the idea behind Body Builders. By following the directions included, you or a parent can help your children construct a model of the human body. The process is simplified by the use of the Body Builder cards found on the next few pages. The cards provide directions and discussion questions for each step. They can be copied, cut apart, and then glued onto large index cards for repeated use. The format is easy to follow and designed so parents or older students can work with one child at a time or a small group of children. When students are finished assembling their bodies, not only will they know the names and functions of their inside parts, they will know that their bodies are amazing.

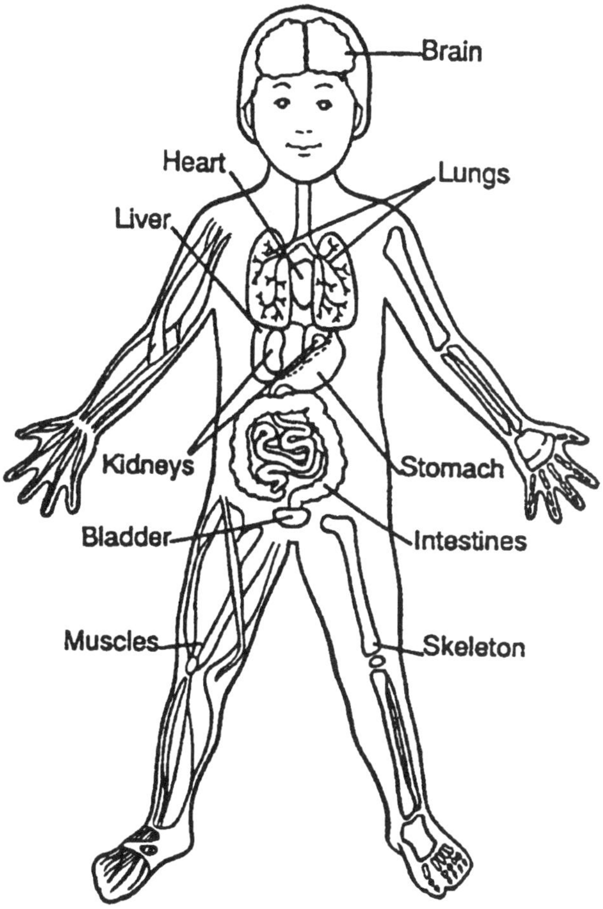

Materials Needed to Complete One Entire Body

- one sheet of butcher paper folded in half (This should be large enough so a child's entire body will fit on each of the folded sides. Standard rolls of 4' (120 cm) paper work well.)
- wide-tipped dark marker
- colored markers or crayons
- mirrors
- paper bones found on pages 63 and 64
- scissors
- glue
- transparent tape
- one sponge
- stapler
- 3.5' (105 cm) of cream colored twine
- 2-20" (50 cm) pieces of cream colored yarn
- 2-25" (62.5 cm) pieces of cream colored yarn
- 15 packing peanuts
- 2 red paper streamers 18" (45 cm) long
- 2 flex straws
- 1 large straw
- plastic bags (sealable and non-sealable)
- 2 small handfuls of fiberfill stuffing
- 4-25" (62.5 cm) pieces of red yarn
- 1-10" (25 cm) piece of red yarn
- 1 small, round red balloon
- 1-13" (32.5 cm) skinny "animal" balloon
- play dough

Body Builder Activity Cards

Body Builder Card 1—The Body Outline

Materials:

- one sturdy sheet of butcher paper folded in half (large enough so child's whole body fits on each side of the folded sheet, standard rolls of 4' [120 cm] paper works well)
- pencil

Directions to complete a body outline:

1. Have one child lie down on the paper and position his/her body as shown.
2. Use a pencil to trace the outline of each child's body on the folded butcher paper. Trace around the child's hair to help individualize each tracing. The fold will be the spine of a life-sized body book, so be sure that each child's right arm lines up against the fold of the paper as shown.
3. After each child stands up, retrace the outline with a marker and add a waistband and collar. Remember to write the child's name on his/her paper.

Discussion Questions:

1. How is your body like your neighbor's?
2. What is the difference between this outline and the "real" you?

Body Builder Card 2—The Front and Back

Materials:

- crayons
- scissors
- mirrors

Directions to complete the front and back of your body outline:

1. Place paper outlines on a large area where children can color.
2. Have each child use a mirror to study his/her features and then color the front of his/her body outline.
3. Cut out their body outlines. Remember that you want the bodies to open like books, so allow the right arm fold to remain intact.
4. After teaching *Back to Back,* have each child color the back side of his/her outline.
5. Hang them up temporarily, with a tack or piece of tape. You will want to bring the outlines down repeatedly as you add the other inside parts.

Discussion Question:

In what ways is your body special?

Body Builder Card 3—The Bones

Materials:

- paper bones, pages 63 and 64
- scissors
- glue or tape

Directions to add paper bones to body outline:

1. Display a picture of the human skeleton.
2. Cut out bones from pages 63 and 64.
3. Glue or tape each bone onto the body outline as shown. The letters on the bones match the letters on the picture above.

Discussion Questions:

1. What jobs do bones have in our bodies?
2. How many bones are in the human body?

Body Builder Activity Cards

Body Builder Card 4—The Brain

Materials:

- a sponge cut to look like a human brain
- scissors
- stapler
- an additional demonstration sponge
- a tub with 1" (2.5 cm) of water

Directions to add the brain to the body outline:

1. Review with your children that the brain takes in information and sends out messages.
2. Have each child place the demonstration sponge in the tub of water to see the absorbtion and then squeeze the sponge to release the water. Explain to them the brain is constantly taking in and sending out information.
3. Staple the precut sponge to the body outline as shown.

Discussion Question:

What protects the brain?

Body Builder Card 5—The Spinal Cord and Nerves

Materials:

- 40" (100 cm) of cream colored twine
- 2-20" (50 cm) and 2-25" (62.5 cm) pieces of cream colored yarn
- stapler
- tape

Directions to add spinal cord and nerves to the body outline:

1. Fold the piece of twine in half and tape the looped end underneath the brain. Allow the excess twine to dangle so the backbone may be added later.
2. Staple or tape nerves (yarn) to the arms and legs. Begin each piece at the spinal cord.
3. Tell your children the bumpy backbones (vertebrae) they feel in their backs will be added later.

Discussion Question:

Why does the spinal cord need to connect to the brain?

Body Builder Card 6—The Backbone

Materials:

- packing peanuts
- tape

Directions to add the backbone to the body outline:

1. Have each child pick a partner and feel each other's backbone. Remind them the backbone protects the spinal cord.
2. Explain to them that there are 34 little bones that protect the spinal cord.
3. Help each child count out 15 peanuts to represent vertebrae.
4. Have them string each peanut as shown. Place a piece of tape at the bottom to keep the peanuts on the string. The backbone should end about waist level.

Discussion Question:

Why do you think the backbone is made up of many bones instead of one long, stiff one?

Body Builder Activity Cards

Body Builder Card 7—The Muscles

Materials:

- 1-8" (20 cm) and 1-10" (25 cm) red paper streamer
- stapler
- glue

Directions to add arm muscles to the body outline:

1. Have each child hold his/her arm straight out and palm up. Have him/her feel the bicep muscle in the upper part of his/her arm. Make a fist and bring the fist up to your shoulder. Can you feel the muscle in your arm bunch up? That is your bicep muscle doing its job.
2. Now, bunch up the 10" (25 cm) streamer and staple it onto the bent arm as shown. Glue the edge of the streamer to the bone and remind them that some muscles are attached to bone.
3. Have each child straighten out his/her arm again to feel the bicep flattening out. Staple the 8" (20 cm) streamer to the straight arm. It should be flat and not bunched. Again, glue the streamer to the bone.

Discussion Question:

When do we use our muscles?

Body Builder Card 8—The Windpipe and Lungs

Materials:

- 2 flex straws
- 1 slightly larger straw
- 2 non-resealable sandwich bags
- 2 small handfuls of fiberfill stuffing
- scissors
- stapler

Directions to add a windpipe and two lungs to the body outline:

1. Cut the two flex straws 2" (5 cm) above the bend.
2. Insert them into the other straw as shown.
3. Staple the top of the straw onto the body. Trim the fat straw if it extends past the neck.
4. Put a small handful of fiberfill into each baggie. Flatten it out and staple the top of each bag to the body as shown.
5. Cut a small hole in each lung and insert the straw.
6. Show your children if they press on the lungs, they can force the air out but it will refill.

Discussion Question:

Our model windpipe does not reach as high as a real windpipe. How high do you think a real windpipe goes?

Body Builder Activity Cards

Body Builder Card 9—The Heart and Circulatory System

Materials:

- spaghetti dyed red (To dye spaghetti, cook pasta until it is soft but still firm. Drain it and mix in red food coloring. No need to glue if used while it is still wet.)
- 1 small, round, red balloon

Directions to add a heart and circulatory system to the body outline:

1. Show your children a picture of the circulatory system.
2. Blow up the balloon and staple the part after the tie to the center of the body outline as shown.
3. Add the red spaghetti, first around the heart and then over the rest of the body.
4. Tell your children the heart pumps blood all over the body through the veins.

Discussion Question:

How does the blood get back to the heart to renew itself?

Body Builder Card 10—Esophagus and Stomach

Materials:

- 1-13" (32.5 cm) thin balloon
- play dough
- one sealable plastic bag
- stapler
- scissors

Directions to add the esophagus and stomach to the body outline:

1. Cut off the closed end of the balloon.
2. Insert the lip of the balloon (esophagus) into the top of the plastic bag (stomach) and seal it. You have made an esophagus and connected it to the stomach.
3. Have children shape a small food item out of play dough. Then have them "eat" the food by tearing it up into tiny pieces.
4. Carefully have children drop the tiny bits of food into the balloon and help them squeeze and push the play dough bits down through the esophagus and into the stomach.
5. Staple the esophagus and the stomach on the child's body outline as shown above.

Discussion Questions:

1. Why was it important to chew your food into very small pieces?
2. How did the play dough food get into the stomach?

Body Builder Card 11—The Intestines

Materials:

- 1 leg of a pair of pantyhose
- stapler

Directions to add the intestines to the body outline:

Begin with one end of the pantyhose leg coming from the stomach and bend, twist, and staple the pantyhose leg onto the body as shown.

Discussion Question:

How will such a long pantyhose leg fit into this small space?

Parent Information

Note: This page should accompany children when they take their body models home.

Your child has been studying the body in school. They have learned that their bodies are amazing biological machines. The model your child brought home today is filled with exciting learning experiences. This sheet will help you and your child to talk about the body.

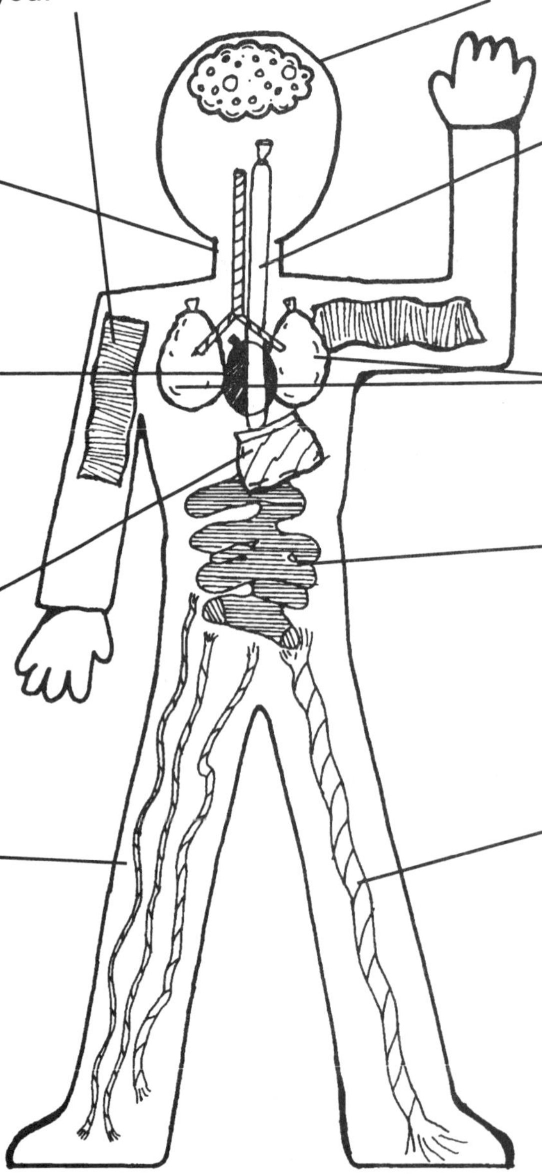

Twisted red streamers represent my bicep muscles. The bicep bunches up when I bend my arm and flattens out when I straighten it. My muscles are attached to the bones in my arm so I can wave hello to you.

Here a sponge is representing my brain. The brain "soaks" up information and is the boss of the body.

This long balloon is my food tube. Food is squeezed through this tube to my stomach.

These straws are really one tube—the windpipe.

These two plastic bags filled with fiberfill are my lungs. The lungs are spongy tissue.

The red balloon represents my strongest muscle—my heart.

Twisted pantyhose represent my intestines. Nutrients are absorbed into the blood here and wastes exit the body.

This sealable plastic bag is my stomach.

The yellow yarn are nerves. Nerves deliver messages from my brain to every part of my body.

The red yarn are blood vessels. Blood moves through the blood vessels to all the part of my body.

Science Award

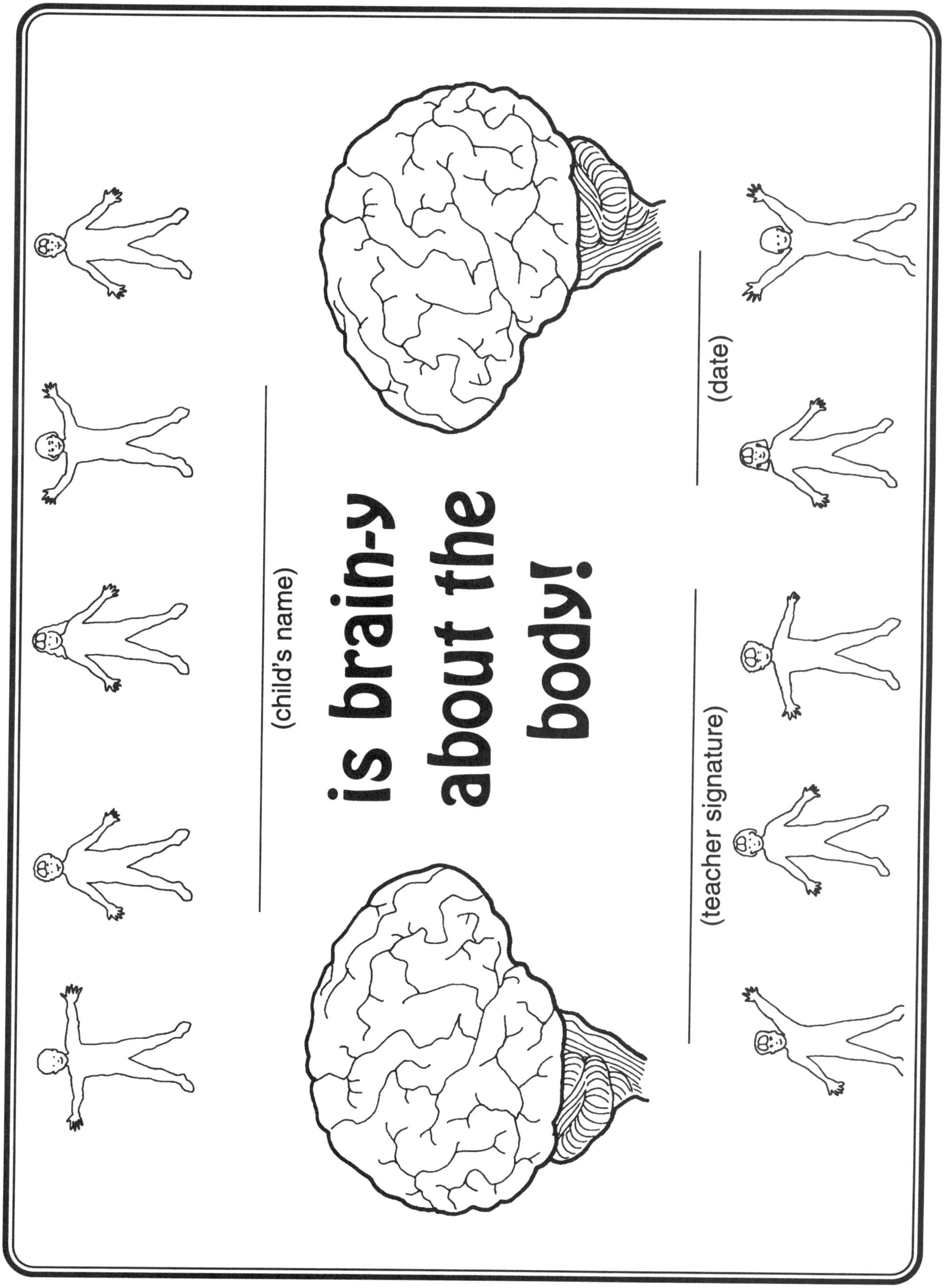

Glossary

B C D E H I L M O P S T W

Blood Vessels—little tunnels that make highways all over your body for the blood to travel.
– tube-like passageways that transport blood throughout the body.

Circulation—blood moving around and around in your body.
– the circulatory system at work moving blood from the heart, in arteries around the body and back to the heart in veins.

Diaphragm—the muscle located under your ribs, helping you breathe.
– thin, strong muscle that separates abdominal organs from the heart and lungs; it contracts and relaxes to deflate and inflate the lungs.

Esophagus—the food tube leading from your mouth to your stomach.
– hollow tube that connects the back of the mouth to the stomach; muscles push food in waves down this tube.

Exhale—to breathe out.
– the action of the ribs moving down and in, the diaphragm relaxing and moving up, and air exiting the body through the mouth and nose.

Heart—a muscle that pumps blood around inside the body.
– muscular organ that lies in the chest between the lungs; the heart has four chambers, two ventricles, and two atria.

Intestine—food goes here after it leaves the stomach.
– long, coiled tube that is divided into the small and the large intestines; food passes from the stomach to the intestines where food is digested; nutrients are absorbed into the body through the intestine walls until only waste remains in the intestine.

Inhale—to breathe in.
– the action of air entering the mouth and nose, the diaphragm contracting and moving down, and the ribs moving up and out.

Lungs—spongy sacks in your chest that take in the air you breathe.
– spongelike sacs that lie on either side of the heart; they contain millions of tiny air sacs where blood picks up oxygen and rids itself of carbon dioxide.

Muscles—tissues under your skin that help you to move.

Oxygen—a gas found in the air we breathe. The body needs oxygen to help it burn the food you eat into body fuel.
– an odorless, colorless gas that is in the air taken in by the body; every body cell needs oxygen.

Posterior—the side of the body that faces backward.

Saliva—a special juice in your mouth that mixes with food when you chew.
– liquid secreted by salivary glands in the mouth to keep the mouth moist and begin the process of digestion.

Smelling Patch—a patch the size of a postage stamp found inside your nose that sends smell signals to your brain.

Taste Buds—tiny tasters on your tongue.
– over 9,000 tiny sense organs found on the tongue, the back of the mouth and the throat; nerve endings in the taste buds detect bitter, sour, salty, and sweet and then pass this information on to the brain.

Torso—the middle section of your body.

Windpipe (trachea)—the tube through which you breathe.
– air passage that leads from mouth to lungs.

Bibliography

Alexander, Sally Hobart. ***Mom Can't See Me.*** MacMillan Publishing Company, 1990.

Aliki. ***My Five Senses.*** Thomas Y. Crowell Company, 1962.

Aliki. ***My Hands.*** Thomas Y. Crowell, 1990.

Balestrino, Philip. ***The Skeleton Inside You.*** Thomas Y. Crowell, 1989.

Carlson, Nancy. ***Arnie and the New Kid.*** Trumpet, 1990.

Carratello, Patricia. ***My Body.*** Teacher Created Materials, 1980.

Cole, Joanna. ***The Magic School Bus Inside the Human Body.*** Scholastic Inc., 1989.

Cole, Joanna. ***Your Insides.*** Putnam and Grosset, 1992.

Crelinsten, Jeffrey. ***To the Limit.*** Harcourt Brace Jovanovich, 1992.

Cumbaa, Stephen. ***The Bones Book and Skeleton.*** Workman Publishers, 1991.

Cummings, Phil. ***Goodness Gracious!*** Orchard Books, 1992.

Fleming, Denise. ***Lunch.*** Henry Holt and Company, 1992.

Fowler, Allan. ***Smelling Things.*** Children's Press, 1991.

Fowler, Allan. ***Tasting Things.*** Children's Press, 1991.

Gomi, Taro. ***Everyone Poops.*** Kane-Miller Books, 1993.

Gossett, Carol, Jill Delano, Virginia Kammer, Valerie Welk and Katy Hall. ***Skeletons! Skeletons! All About Bones.*** Putnam Publishing Group, 1991.

Hoban, Tana. ***Exactly the Opposite.*** Trumpet Club, 1990.

Nichols, Breck. ***My Body.*** Teacher Created Materials, Inc., 1992.

Nikola-Lisa, W. ***Bein' with You This Way.*** Lee & Low Books, 1994.

Parker, Steve. ***Food and Digestion.*** Watts, 1990.

Parker, Steve. ***How the Body Works.*** Dorling Kindersley, 1994.

Parker, Steve. ***Touching a Nerve: How You Touch, Sense and Feel.*** Franklin Watts, 1992.

Rankin, Laura. ***The Handmade Alphabet.*** Scholastic, 1991.

Richardson, James. ***Science Dictionary of the Human Body.*** RGA Publishing Group, 1992.

Rius, Maria, J.M. Parramon, and J.J. Puig. ***Touch, Sight, Taste, Smell, and Hearing.*** Barron's, 1985.

Royston, Angela. ***What's Inside My Body?*** Dorling Kindersley, 1991.

Sainsbury, Mal, Merilyn Holme, and Josephine Paker. ***Human Body.*** Merlion Publishing Ltd., 1992.

Showers, Paul. ***Ears Are for Hearing.*** Thomas Y. Crowell, 1990.

Showers, Paul. ***How You Talk.*** HarperTrophy, 1992.

Showers, Paul. ***Your Skin and Mine.*** HarperTrophy, 1991.

Simon, Norma. ***Why Am I Different?*** Albert Whitman, 1993.

Stein, Sara. ***The Body Book.*** Workman Publishing, 1992.

Suzuki, David. ***Looking at the Body.*** John Wiley & Sons, Inc., 1991.

Wessels, Florence and Pierre-Oliver. ***My Body.*** Random House, 1992.

Wood, Carol. ***Sense-able Science.*** AIMS Education Foundation, 1994.

Bibliography *(cont.)*

Technology

Broderbund. ***A.D.A.M. Essentials.*** Available from Learning Services, (800)877-9378. CD-ROM

CLEARVUE. ***The Human Body.*** Available from Educational Resources, (800)624-2926. CD-ROM

DK. ***The Ultimate Human Body.*** Available from Learning Services, (800)877-9378. CD-ROM

Hartley. ***Forgetful Freddy.*** Available from Educational Resources, (800)624-2926. CD-ROM

IVI Publishing. ***AnnaTommy from Mayo Clinic.*** Available from Educational Resources, (800)624-2926. CD-ROM

IVI Publishing. ***Mayo Clinic Family Health Book.*** Available from Educational Resources, (800)624-2926. CD-ROM

IVI Publishing. ***Mayo Clinic the Total Heart.*** Available from Educational Resources, (800)624-2926. CD-ROM

IVI Publishing. ***Welcome to Bodyland.*** Available from Educational Resources, (800)624-2926. CD-ROM

IVI Publishing. ***What Is a Belly Button?*** Available from Educational Resources, (800)624-2926. CD-ROM

Knowledge Adventure. ***3D Body Adventure.*** Available from Educational Resources, (800)624-2926. disk and CD-ROM

Knowledge Connections. ***Body and Mind.*** Available from Learning Services, (800)877-9378. CD-ROM

MECC. ***Body Scope.*** Available from Learning Services, (800)877-9378. CD-ROM

Optimum Resource. ***Top Class Learning Collection: Learn About the Human Body.*** Available from Educational Resources, (800)624-2926. CD-ROM

Queue. ***Learning All About Cells and Biology.*** Available from Learning Services, (800)877-9378. CD-ROM

Tom Snyder Productions. ***Choices, Choices: Kids and the Environment, Choices, Choices: On the Playground, Choices, Choices: Taking Responsibility,*** and ***Inner Body Works.*** Available from Educational Resources, (800)624-2926. disk

StarPress Multimedia. ***Dr. Health'nstein's Body Fun.*** Available from Educational Resources, (800)624-2926. CD-ROM

Virtual Entertainment. ***Body Park.*** Available from Educational Resources, (800)624-2926. disk

